Protecting Chips Against Hold Time Violations Due to Variability

Gustavo Neuberger · Gilson Wirth
Ricardo Reis

Protecting Chips Against Hold Time Violations Due to Variability

Gustavo Neuberger
Instituto Federal de Educação
Ciência e Tecnologia do Rio Grande
do Sul(IFRS) Campus
Canoas Canoas, Rio Grande do Sul, Brazil

Ricardo Reis
Instituto de Informática
Universidade Federal do Rio Grande
do Sul (UFRGS)
Porto Alegre, Brazil

Gilson Wirth
Depto de Eng Elétrica
Universidade Federal do Rio Grande
do Sul (UFRGS)
Porto Alegre, Brazil

ISBN 978-94-017-7794-0 ISBN 978-94-007-2427-3 (eBook)
DOI 10.1007/978-94-007-2427-3
Springer New York Dordrecht Heidelberg London

Printed on acid-free paper

Springer is part of Springer Science+Business Media (www.springer.com)

To my parents Lotário and Isabel, who always gave me the support to follow my dreams.

"Gustavo Neuberger"

To my wife Mariana, who always gave me unconditional love and support.

"Gilson Wirth"

To my wife Lucia, to my daughter Mariana, to my sons Guilherme and Eduardo, and to my parents Constantino and Maria de Lourdes.

"Ricardo Reis"

Contents

About the Authors

Gustavo Neuberger Instituto Federal de Educação, Ciência e Tecnologia do Rio Grande do Sul (IFRS) Campus Canoas Canoas, Rio Grande do Sul, Brazil

Ricardo Augusto da Luz Reis Professor at Universidade Federal do Rio Grande do Sul (UFRGS), Instituto de Informática, Graduate Program on Computer Science (PPGC) and Graduate Program on Microelectronics (PGMicro), Porto Alegre, RS, Brazil

Gilson Inácio Wirth Departamento Engenharia Elétrica and Graduate Program on Microelectronics (PGMicro), Professor at Universidade Federal do Rio Grande do Sul (UFRGS), Escola de Engenharia, Porto Alegre, RS, Brazil

Chapter 1
Introduction

Process Variations and Flip-Flops

With the development of Very-Deep Sub-Micron (VDSM) technologies, process variability is becoming increasingly important and is a very important issue in the design of complex circuits. Process variability is the statistical variation of process parameters, meaning that these parameters do not have always the same value, but become a random variable, with a given mean value and standard deviation. This effect can lead to several issues in digital circuit design.

The logical consequence of this parameter variation is that circuit characteristics, as delay and power, also become random variables. Because of the delay variability, not all circuits will now have the same performance, but some will be faster and some slower. However, the slowest circuits may be so slow that they will not be appropriate for sale. On the other hand, the fastest circuits that could be sold for a higher price can be very leaky, and also not very appropriate for sale. A main consequence of power variability is that the power consumption of some circuits will be different than expected, reducing reliability, average life expectancy and warranty of products. Sometimes the circuits will not work at all, due to reasons associated with process variations. At the end, these effects result in lower yield and lower profitability.

To understand these effects, it is necessary to study the consequences of variability in several aspects of circuit design, like logic gates, storage elements, clock distribution, and any other that can be affected by process variations. The main focus of this book will be storage elements.

Modern synchronous digital designs necessarily include a large amount of flip-flops (FF) in pipeline stages to improve data throughput. FF timing is determined by the CLK-Q propagation time, setup time and hold time. The setup time is the amount of time that the data input must be ready before the clock edge, while the hold time is the amount of time the flip-flop input data must remain stable after the clock edge. The variation of the propagation time due to process variability has been already investigated by Monte Carlo simulation [12]. While statistical variations of setup and propagation times in critical paths are essential for maximum chip performance, a violation of the hold time in short FF-logic-FF paths leads to a chip failure due to a generation of races in the pipeline. Race conditions are caused

G. Neuberger et al., *Protecting Chips Against Hold Time Violations Due to Variability*,
DOI 10.1007/978-94-007-2427-3_1,

by the combination of short paths, clock skew, and jitter between sending and receiving FFs, and process variations. The internal race immunity is a figure of merit to characterize the robustness of a FF against race conditions and is defined as the difference between clock-to-Q delay and hold time. Hence, the race immunity can vary greatly between different FF types [30].

Since modern digital CMOS designs, such as microprocessors, DSP cores, and dedicated hardware accelerators typically comprise thousands of FFs, a statistical analysis of the internal race immunity in combination with clock uncertainties is mandatory. Especially scan chains for DFT schemes [23], where FFs are connected in a serial fashion to build up a shift register during test mode, are sensitive since no logic is placed between the FFs. Therefore, several techniques for diagnosis of single or even multiple hold time failures in scan chains are proposed [13, 20, 23, 26]. There are also techniques to diagnose these failures in generic short logic paths [51] and buffer insertion to increase the delay of these paths. For example, hold time fixing, or padding, is typically done during chip design [41]. However, depending on the design and FF properties, without detailed analysis of the critical clock skew and process variability, the extra delay introduced during hold-time fixing can be over or under estimated.

In this book, we therefore present a statistical analysis of the critical clock skew in several test paths, due to process variability in 130 and 90 nm CMOS technology. To facilitate an on-wafer test, a measurement circuit with a precision compatible to the speed of the technology is needed.

Once measured data is made available, several analyses may be performed on the data. Mathematical methods will be used to separate the variability between systematic and random residual, since each one of them can have different effects and different methods to cope with them.

One issue that must be addressed is whether the measured distributions are normal Gaussian distributions or not. Different normality tests are available to check if this assumption is true. Although most process parameter distributions are known to be normal Gaussian, it is not necessarily true that higher-level measures (delay, hold time, etc) will be Gaussian also.

Another important task to see the relevance of the issue of hold time violations is to combine the data on expected clock skews available in the literature with our data on FF race immunity. A circuit with a small clock skew will probably not pose problems regarding hold time violations, no matter how small the race immunity is; it is also true that a large clock skew can be catastrophic to this issue even if the FFs have large race immunity. So, the probability of hold time violations considering all the data available about race immunity and clock skew must be evaluated.

With all the data and analyses available, it is possible to propose methods to protect digital circuits against these issues, and to check if the empirical techniques used in the industry are in fact the most effective ones.

This book is organized as follows. Chapter 2 discusses the basic aspects and causes of process variability. Chapter 3 presents the FF characterization, and shows the impact of FF parameters on the hold time violation probability. In Chap. 4, we show the circuits that will be used as test cases. Chapter 5 contains the extra circuits needed to perform on-chip measurement. Chapter 6 presents the measurement results already obtained that will be the basis to further analysis. The separation of variability between systematic and random residual, together with the mathematical methods used to perform this task, is addressed in Chap. 7. The normality tests generally used to check if a distribution is a normal Gaussian distribution, and the results achieved with our data are presented in Chap. 8. Chapter 9 shows the calculation of probability of hold time violations when combining our measured data with typical clock skews reported in literature. Chapter 10 presents the options available to protect digital circuits against hold time violations, both after and before circuit fabrication. Finally, in the Chap. 11, we present the final remarks of this book.

Chapter 2
Process Variability

Overview

Parameter variability has always been an issue in integrated circuits. However, comparing with the size of devices, it is relatively increasing with technology evolution, as the device size shrinks in a larger scale than our control over them. Also, in the past, the variations were mostly due to imperfect process control, but now intrinsic atomistic variations become more important, as devices of atomic sizes are achieved. This parameter variation causes uncertainties in circuit design, as in timing, power dissipation, and others important properties. Figure 2.1 shows the technology scaling, to exemplify how small the devices are becoming. Approaching the atomic scale is very difficult to control the process, as only one atom can make a huge difference.

On the other hand, new research from Intel at [25], show that variability is not an insurmountable barrier to technology development, but just a new challenge to overcome and cope with in technology fabrication and design. They successfully showed that it is possible to remain variability under control up to 45 nm, and presented design mitigation techniques. In their experiments, systematic variability was constant at about 3 % of the maximum frequency, and random variability increased from 130 to 90 nm and 65 nm at a maximum of 2.5 %, before decreasing to about 1 % at 45 nm, when they introduced new fabrication control.

This chapter will address the main aspects of process variability. First, we will identify the different sources and types of variations. Then we enter in specific issues that cause variations, as lithography and doping. Finally, we must analyze the true impact on circuit design.

2.1 Sources and Types of Variations

Variations can be classified in many different ways. For example, they can be divided in different sources and types, as in [17]; or between systematic or random sources of variations, as in [39].

The variations can have different sources and types, as presented in [17]. The sources can be divided between process variations, environment variations and

G. Neuberger et al., *Protecting Chips Against Hold Time Violations Due to Variability*,
DOI 10.1007/978-94-007-2427-3_2,

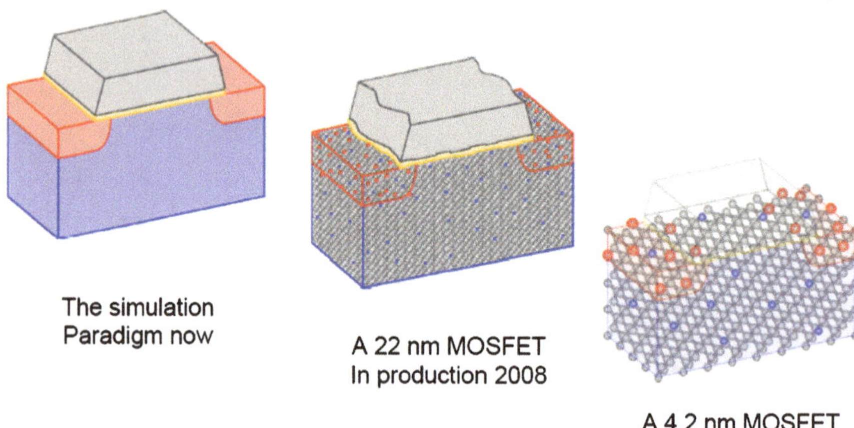

Fig. 2.1 Expected technology scaling [3]

Table 2.1 Matrix of variations

	Process	Environment	Temporal
Global	<L_g>and<W>, <layer thicknesses>, <R>'s, <doping>, <t_{ox}>, <V_{body}>	Operating temperature range, V_{DD} range	<NBTI>and hot electron shifts
Local	Line Edge Roughness (LER), discrete doping, discrete oxide thickness, R and V_{body} distributions	Self-heating, hot spots, IR drops	Distribution of NBTI, voltage noise, SOI V_{body} history effects, oxide breakdown currents

temporal variations. Process variations are variations due to lack of control on the fabrication process, since no two devices are exactly the same at atomic level. Environment variations are variations due to lack of perfect control over the environment (temperature, voltage, etc) in which the circuit must operate. Finally, temporal variations are variations, which cause the device to behave differently at different times, as NBTI, for example.

The types of variations can be divided in global and local. Global are variations in the value of a parameter for the entire wafer, while local are device-to-device variations within any single chip. This distinction is important because these types require different statistical treatment for proper determination of impact on yield.

With these two classifications, we can build a matrix of variations, as shown in [17]. This matrix is presented in Table 2.1.

The main focus here is about process variations, both global and local effects. In the next section we will discuss specific issues about process variations.

Figure 2.2 shows the classification of variations employed in [39]. The variations are divided into two main categories: systematic and random variations. This classification is motivated by the differences in the root causes of the different types of variation.

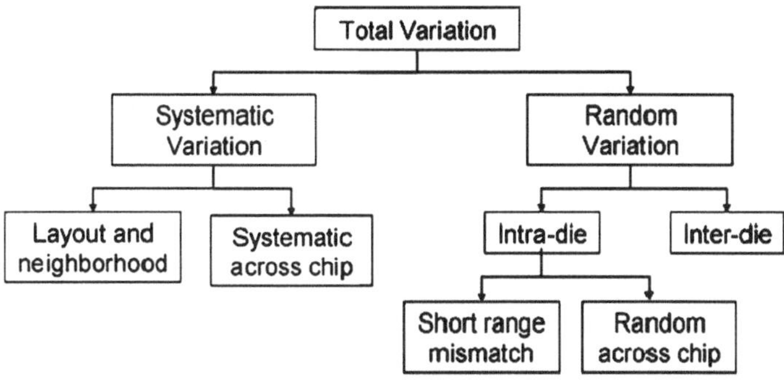

Fig. 2.2 Classification of the different types of variations in transistor characteristics [39]

According to [39], systematic variation is the difference in the electrical characteristics of two transistors with identical width and length (W/L), where there is a clearly identifiable difference in either the device layout or the layout neighborhood. Examples include the impact of gate poly pitch on gate length due to optical proximity effects, stress effects, orientation effects, etc. Random variation is the difference in the electrical characteristics of devices with identical geometry (W/L), layout, and neighborhood within the interaction distance of the known sources of variation. For example, the impact of gate polysilicon pitch differences becomes negligible for pitches greater than 3λ/NA [52], where λ is the wavelength, and NA is the numerical aperture of the exposure system. For λ = 193 nm and NA = 0.75, these give an interaction distance of approximately 775 nm. This forms a lower bound on the distance over which the local neighborhood is required to be identical for the variation to be considered random. Recently, a new source of variation with millimeter-scale interaction distance has been observed: rapid thermal anneal (RTA) temperature variation due to the density of STI regions not covered by gate poly [43, 2]. Such extreme long interaction-distance effects are considered to be random variations by [39].

Each of the main categories can be further divided. Classification of random variation into intradie (within-die) and interdie variations (between different dies) helps in identifying the root causes and possible improvement actions. For example, intradie variation can be addressed by lithography improvements, such as off-axis illumination [27] and within-field-exposure dose compensation [48]. Across-wafer nonuniformity's contribution to interdie variation can be improved by techniques like exposure dose compensation across the wafer [48], ion-implantation uniformity, RTA uniformity, etc. Systematic variation can be divided into variation arising from layout or neighborhood differences within an interaction distance.

Different aspects of circuit fabrication cause the process parameter variations. The two major sources of process variations are the fabrications steps of lithography and doping. These two will be discussed in detail.

Table 2.2 Shot noise for different energy quanta

Lithography	Energy (eV)	Resist dose (mJ/cm²)	# quanta per 50 nm pixel	3σ dose variation (%)
193 nm	6.4	20	500,000	0.4
EUV – 13.5 nm	92	2	3,400	5
X-Ray – 1.3 nm	920	40	6,800	4
E-beam	50,000	150 (3 μC/cm²)	470	14
Ion-beam	100,000	50 (0.5 μC/cm²)	78	34

2.1.1 Lithography

The lithographic sources of variations are the cause of both global and local variations. Imperfect lithographic process control and errors in alignment, rotation and magnification are the problems that lead to global variations. The critical dimensions are sensitive to focus, dose (intensity and time), resist sensitivity (chemical variations) and layer thicknesses [4, 9].

The local lithographic variations are due to pattern sensitivity (interference effects from neighboring shapes), interference effects from buried features, and LER (Line Edge Roughness) [10].

LER, which is primarily a discreteness effect, is due to sources of statistical variation in chemically amplified resists. These variations include fluctuations in the total dose due to finite number of quanta (shot noise), fluctuations in the photon absorption positions, nanoscale non-uniformities in the resist composition, statistical variations in the extend of acid-catalyzed deprotection, and statistical effects in polymer chain dissolution. For example, Table 2.2 shows the estimated dose uncertainty for a 50 nm contact hole, with different lithographic processes.

2.1.2 Doping

The doping process causes mainly local variations, as there are less and less dopant atoms in the transistor channel, in every new technology node. Working with few atoms can lead to a strong variability in the threshold voltage, which is proportional to the square root of the number of dopant atoms.

As showed in [15], the number of dopant atoms in the depletion layer of a MOSFET has been scaling roughly as $L_{eff}^{1.5}$. Statistical variation in the number of dopants, N, varies as $N^{-1/2}$, increasing V_T uncertainty for small N. And also, specific V_T uncertainties depend on the details of the doping profiles. Figure 2.3 shows a graphic of the scaling of number of dopant atoms versus scaling.

This doping uncertainty has a huge impact in the threshold voltage variability. Frank [16] shows an experiment with V_T measurements on 3481 identical SOI nFETs, all of single experimental macro on a single wafer (Fig. 2.4). The standard deviation can be up to 10 % of the mean threshold value.

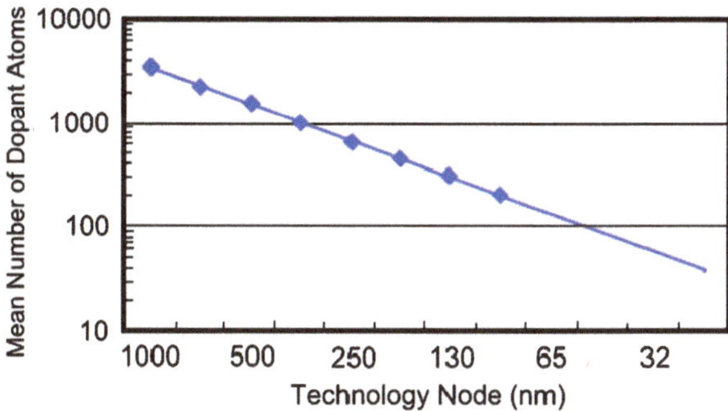

Fig. 2.3 Number of dopant atoms versus scaling [8]

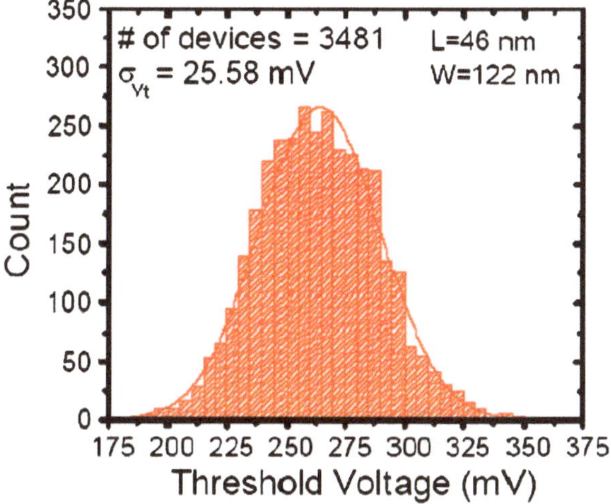

Fig. 2.4 Threshold variability [15]

2.2 Impact on Circuit Design

Process variations and intrinsic device variations cause logic and memory yield loss. Since the variations are Gaussian, and not bounded, it is not possible to absolutely guarantee functionality. Therefore, circuit designers must learn how to cope with variability, and the design must be based on achieving a target yield (90 %, 98 %, etc). The design automation tools must have means to evaluate correctly the yield.

Today, EDA tools evaluate the designs using corners. Designers usually simulate using nominal parameter values, worst-case values, and best case values. And they attempt to achieve high yield at the worst case. This is clearly too pessimistic, as

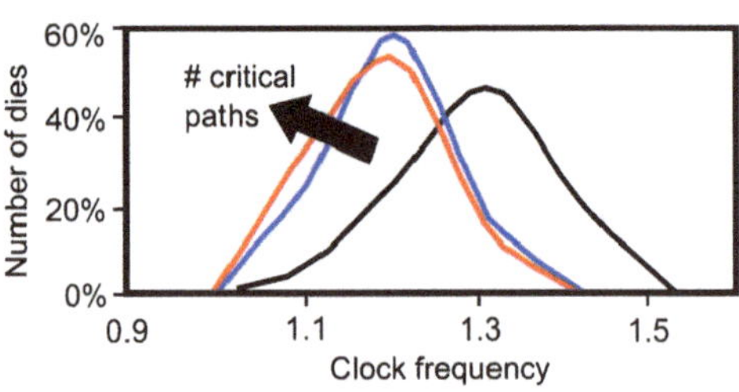

Fig. 2.5 Impact of number of critical paths on the frequency distribution [8]

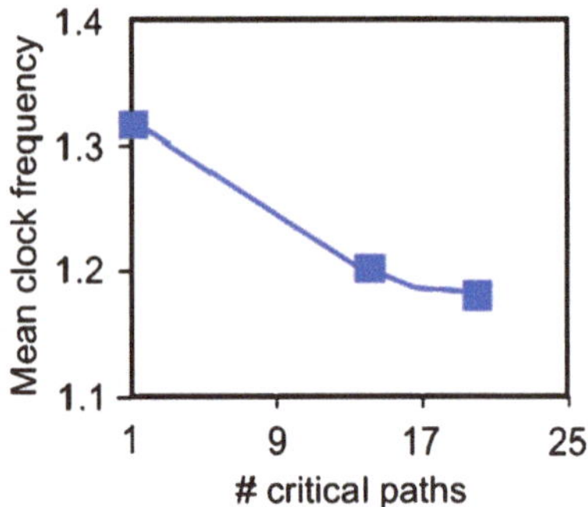

Fig. 2.6 Impact of number of critical paths on the mean frequency [8]

with process variations, it is very rare that all the devices will simultaneously show this worst-case value. It is much more possible that some of them go to one direction and others to another. This leads to statistical timing analysis tools. Now, the paths do not have one deterministic timing result, but one statistical result.

According to [37], it is necessary to verify the product response to all variations. The main responses are in power and frequency. Power variability includes variation in IDDQ (Vth fluctuation, temperature spread, voltage drop) and AC power spread (device capacitance, metal capacitance), while frequency variability includes Lg variations, thermal variations, voltage variations, and metal thickness variations. Another important product response is the possibility of failure due to setup time and hold time violations, which are a consequence of delay variability and clock skew.

In general, the fabricated circuits will show a Gaussian range of performance. Some will be so slow that will not be sold, while others will be very fast, but also too leaky. The ones near the mean value will be the good ones that go to the market.

Another serious issue related to variability and circuit design is the performance loss with the increase of the number of critical paths. As they are uncorrelated, with many critical paths, the probability that only one becomes slower than the nominal case is larger. Figures 2.5 and 2.6 shows the impact of the number of critical paths on performance.

These are the main problems that can be seen in circuit design due to process variability. It is important to develop techniques to cope with these problems, to achieve better performance and higher yield, increasing profit.

2.3 Design Techniques with Process Variations

In recent technologies, the circuits are very susceptible to yield loss. In order to cope with this loss, the design must use techniques to address yield. Currently, with the development of technology, the design challenge is changing from defect-limited yield to parametric-limited yield, which is the effect of process variations.

Since the variations are Gaussian, and not bounded, it is not possible to absolutely guarantee functionality. Therefore design must be based on achieving a target yield (e. g., 90 %). But how to evaluate the expected yield is a design automation challenge. However, if this is not taken into account in the design, there will be a loss of performance, which increases in each technology generation, as shown in Fig. 2.7. So, sooner or later, it will be necessary that all designs must be variability-aware.

This section will present three different possible design techniques to cope with process variations: Monte Carlo simulations, corner analysis, and statistical static timing analysis.

2.3.1 Monte Carlo Simulations

Monte Carlo simulation is the simplest method to predict yield, but it is also the most time consuming. It is based on a series of hundreds of normal simulations, with each simulation randomly using different electrical parameters for the transistors and the other elements that are subject to variations. These parameters can vary in a Gaussian form, or also in any other appropriate form.

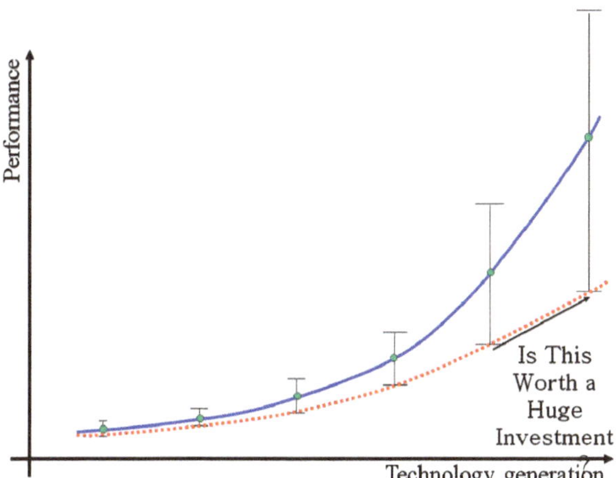

Fig. 2.7 The march of technology

After the Monte Carlo simulation is complete, it is necessary to analyze the results, and calculate the percentage of simulations that achieved the desired results, and that is the yield. However, if a high enough yield is not achieved, it is necessary to change the design, and repeat all the simulations again.

Since the simulations are very time consuming, this method can not be applied to very large designs, only to designs with a few dozen gates. However, it can also be useful to use it to single gates, and use the results in the following methods: corner analysis and statistical static timing analysis.

2.3.2 Corner Analysis

Corner analysis is a traditional approach that ensures good yield, at the expense of a pessimistic design, and also it is not possible to know the distribution of the yield or performance, only that the yield will be very high. A detailed explanation of corner analysis can be found in [31].

In the corner analysis, all electrical parameters are considered to be within certain bounds min and max. When they are Gaussian distributions, these bounds are normally considered to be ±3-sigma. If, during chip design, one finds that the circuit meets the performance constraints for all corners (boundaries), then the design is considered acceptable. Corner analysis has been applied for many years to design chips that are robust to variability.

In the corner-based analysis method, the circuits are design so that the functionality is guaranteed at all worst-case corners. The values of parameters are considered deterministic, and no probability is taken into consideration. However, the drawbacks of this method are that it makes sense only if sources of variations are strongly correlated (worst case occurs for all parameters), and it is clearly too pessimistic for deep sub-micron technologies, where there are more random variations than systematic variations.

2.3.3 Statistical Static Timing Analysis (SSTA)

With the increasing variability in the manufacturing process, corner analysis has been considered as inadequate and overhaul, and statistical analysis has been proposed as an alternative approach. Specifically statistical static timing analysis (SSTA) has been proposed as an alternative to traditional static timing analysis (STA) [31].

Most SSTA methods proposed use propagation of distributions of delay through the logic network. Considering the signal arrival time as a Gaussian random variable, the method consists in the propagation of the pdf (probability density function) of the random variables through the logic network, as illustrated by Fig. 2.8. Path-based SSTA does this for a given path, while block-based SSTA generates the pdf for the maximum delay of the block [31].

- Deterministic:

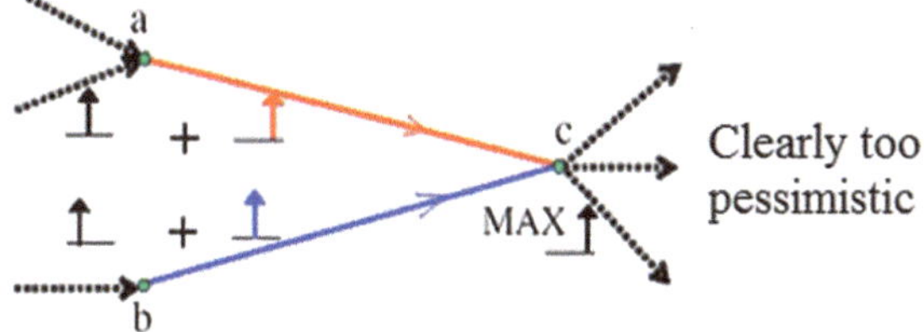

- Statistical:

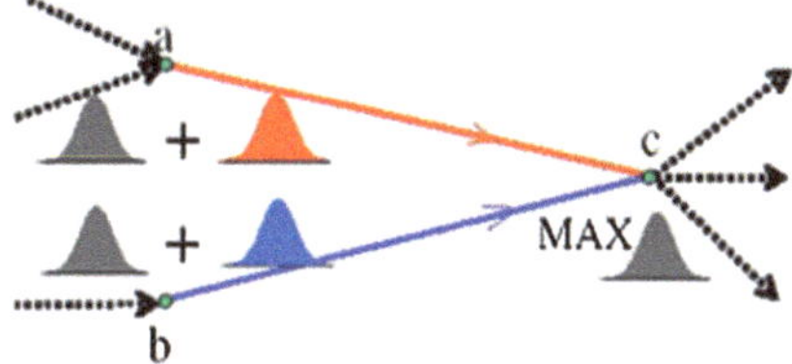

Fig. 2.8 Block-based Statistical Timing Analysis

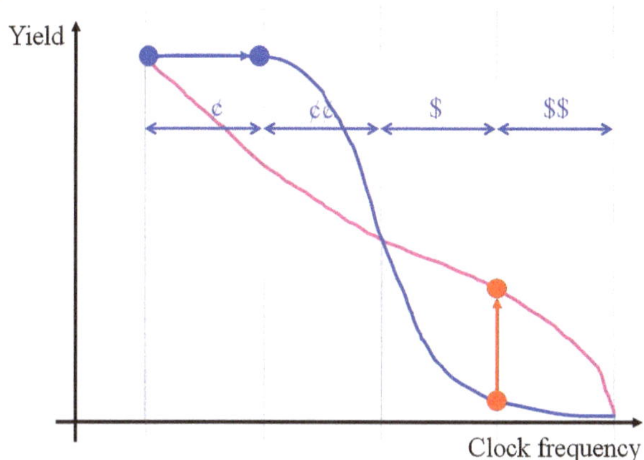

Fig. 2.9 Parametric yield curves showing profitability

The current available SSTA tools propagate the Gaussian random variables, and the final results can be used to determine the yield. Knowing the final distribution can be very useful to analyze different scenarios and determine which one is more appropriate to maximize yield (or profitability). For instance, consider Fig. 2.9. One curve is corresponding to a design with a narrow distribution of final performance. It has a high yield, since not many circuits are so slow that cannot be sold for a profit. However, it also does not have dies that will be very fast, and that can be sold for a much higher profit. When considering a wider distribution, a more significant number of dies will have to be discarded, but also many more dies will be very fast and sold by a very higher price, and the final result will be a higher profit

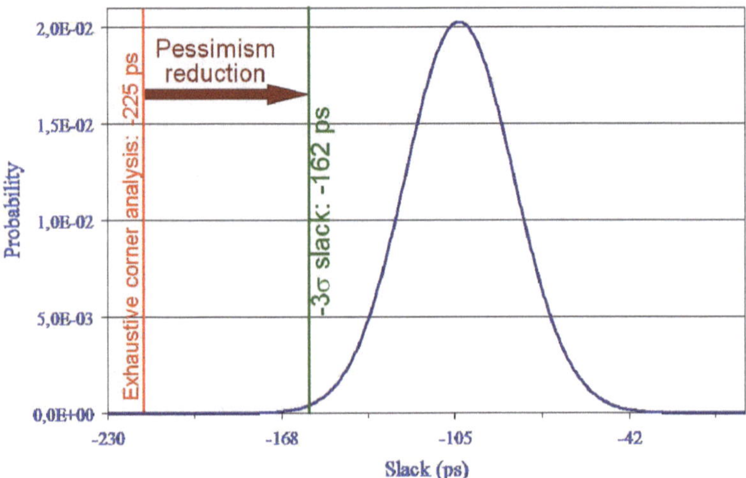

Fig. 2.10 Corner-based versus statistical timing on ASIC part

for the company. So, it is not easy to know which distribution is more appropriate for each case.

There are different tools showing the advantages of SSTA over corner analysis. For example, [49] obtains almost a 30 % improvement in the slack prediction over corner analysis, with the added advantage of knowing the exact final distribution, as presented in Fig. 2.10.

Chapter 3
Flip-Flops and Hold Time Violations

Definitions

Digital circuits necessarily include storage elements to ensure correct operation. The most common are the flip-flops (FFs). Although they are not very difficult to use, their operation must be understood and they must be characterized due to their timing metrics. They are not immune to failures, and one failure that FFs can present is the hold time violation, if they are not carefully designed. This chapter discusses FF operation and metrics, and hold time violations.

3.1 Flip-Flop Timing Metrics

FFs are edge-triggered clocked storage elements. They are frequently built from latches [29]. They connect the input data to the output data in the clock's rising (or falling) edge, holding the signal stable until the next rising (falling) edge. FFs are usually used to store data during a short or long time and also to store data only during a pipeline cycle of a circuit. Transitions at the FF input do not propagate to its output, unless during the edge of the clock cycle. However, to ensure correct operation, FF data and clock input must satisfy some timing restriction.

As defined in [29], the FF timing metrics can be physical: setup time, hold time, clock-to-output delay and data-to-output delay; or synthetic, built to help system design: delay and internal race immunity. This section will detail all of them.

3.1.1 Setup Time

The common definition of setup time is the time interval before the active clock edge during which data is not allowed to change, in order to ensure FF correct operation [11, 22, 36]. However, this classical definition is not very precise, because there is an increase in clock-to-output delay when data arrives very close to setup time.

G. Neuberger et al., *Protecting Chips Against Hold Time Violations Due to Variability*,
DOI 10.1007/978-94-007-2427-3_3, © Springer Science+Business Media Dordrecht 2014

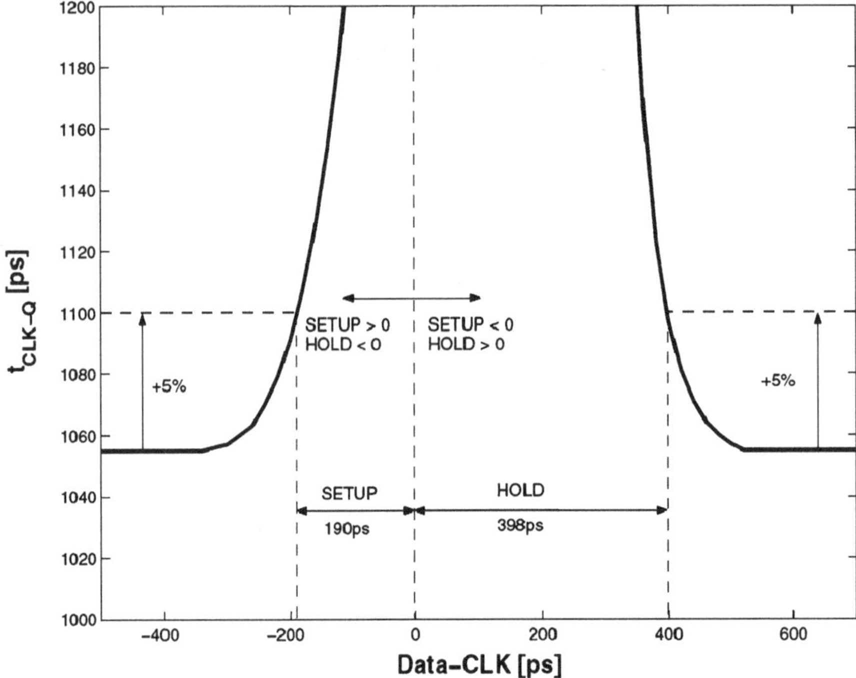

Fig. 3.1 Definitions of setup and hold times [29]

An alternative definition for setup time, frequently used by ASIC standard cell libraries, is based in the fact that a 5 % increase from the nominal value in the clock-to-output delay can be tolerated. With this assumption, [29] makes the following definition:

> Setup time is the minimum data-to-clock offset that causes the clock-to-output delay to be 5 % higher than its nominal value.
>
> (1) Setup of logic '1' is the setup time measured when data undergoes a 0-1 transition.
> (2) Setup of logic '0' is the setup time measured when data undergoes a 1-0 transition.
>
> Setup time can be positive or negative, depending especially on the FF circuit topology. Figure 3.1 shows the graphical definition of setup time.

3.1.2 *Hold Time*

Hold time is analogous to the setup time, but measured after the active clock edge. In [29], hold time is also defined based on the tolerance to an increase in the delay:

> Hold time is the minimum time interval during which a new data has to retain its value after the active clock edge so that clock-to-output delay is 5 % higher than its nominal value.
>
> (1) Hold of logic '1' is the hold time measured when data undergoes a 0-1 transition.
> (2) Hold of logic '0' is the hold time measured when data undergoes a 1-0 transition.

If data changes too fast, the circuit may not work correctly. As setup time, hold times can also be positive or negative, depending mainly on circuit topology.

3.1.3 Clock-to-Output Delay

According to [29], the clock-to-output delay is the delay measured from the active clock edge to the output. It depends on the data-to-clock offset, clock slope, supply voltage, and output load. The rising and falling delays cannot, in general, be identical because the two cases will experience a variety of rising and falling delays [21].

3.1.4 Data-to-Output Delay

The last physical metric is the data-to-output delay. It is the delay measured from a 0-1 or 1-0 data transition to the output, assuming that the FF is clocked correctly. The data-to-output delay is not a good metric of a FF performance because it depends on the arrival of new input data relative to clock [29].

3.1.5 Delay

For high-level system design, it is important to have one relevant metric to describe the true impact of the FF in the circuit. It is possible to reduce the metric to one FF delay parameter with respect to system performance [46]. It is defined as following:

> The delay of a FF is the sum of its clock-to-output delay (measured at setup time) and setup time.

This definition is independent on simulation setup, avoiding misconceptions present in literature about FF speed, often characterized as clock-to-output delay [5, 47].

3.1.6 Internal Race Immunity

The synchronous digital systems are subjected to clock skew, which is the maximum difference between two different FF clock inputs. So, maximal clock skew that a system can tolerate is determined by FFs. To quantify this FF timing metric, the concept of internal race immunity is introduced [29]:

> Internal race immunity of a FF is the difference between its clock-to-output delay and hold time.

Equation 3.1 gives the formula of this definition.

$$R = t_{CLK-O} - t_{hold} \tag{3.1}$$

This is a helpful metric to help designers to prevent against timing failures due to short-paths (races). It also tells the maximum clock skew that a FF can tolerate. However, different FFs have different race immunities. Typically, faster FFs have smaller race immunity, and slower FFs, such as the master–slave FF, have larger race immunity.

3.2 Hold Time Violations

The most common source of timing failures in high-speed digital systems is the lack of race immunity (race immunity is smaller than the clock skew). It is usually named hold time violations, or race conditions.

Considering two logically adjacent FFs, controlled by CLK1 and CLK2, with no logic between them, they are potentially subjected to a clock skew. If the clock skew is large enough, i.e. CLK2 arrives after CLK1 and exceeds the internal race immunity $t_{CLK-Q} - t_{hold}$ of the FF, a race is produced and detected if the output of both FFs are of same value at same time (Q1(t) = Q2(t)). The violation can be detected by initializing the FFs with opposite values, and applying a pulse in the data input, as shown in Fig. 3.2. As long as Q1(t) ≠ Q2(t) pipeline operation is correct.

Equation 3.2 describes the timing conditions in the case of a violation. Especially, fast FFs with large hold times are sensitive to hold time violations. Δt_{var} includes variations from different sources.

$$t_{CLK-Q} - t_{hold} - t_{CLKskew} - \Delta t_{var} < 0 \tag{3.2}$$

It is possible to see that the FF race immunity (that is inherent to the FF type and transistor sizes used in the design), the maximum clock skew found in the circuit,

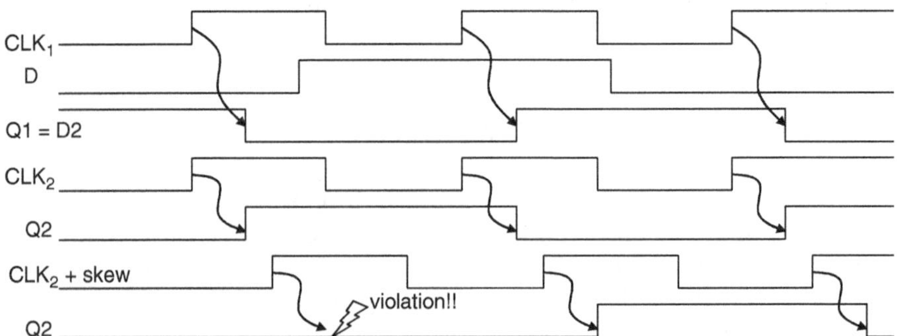

Fig. 3.2 Timing diagram showing hold time violation

and process variations contribute to the probability of a hold time violation. If the clock uncertainty is very well controlled and race immunity is large enough, process variability plays a minor role, but this is not the case of the majority of semi-custom designs that have to meet short time-to-market. Usually, the clock uncertainty and race immunity are of about the same order of magnitude, as it is going to be presented later on this text.

Chapter 4
Circuits Under Test

For Characterization

The circuits can be very sensitive to process variability, but different circuits can have different sensibilities. To have representative results, the circuits that will be fabricated to be tested must be carefully chosen. First, the sensitivity of the logic circuits will be verified through simulation in 130 and 90 nm technologies, however only the results for 90 nm will be shown in this chapter. Using MC simulations, the sensitivity of inverters (representing generic combinational circuits) and FFs will be measured. Combining them, we will verify if hold time violations are a potential problem, and if they can be generated by process variations. Finally, the circuits chosen for fabrication and measurement in silicon will be shown.

4.1 Inverter Sensitivity

We know from the process specifications the expected variability of the process parameters. However, it does not tell about the true effect in the delay variability of the logic circuits. The sensitivity of these logic circuits must be analyzed to find the potential problem. A chain of two inverters will be used to investigate it (Fig. 4.1). Monte Carlo simulations with 100 iterations were performed to estimate both the global and local variations. Different inverter sizes were also analyzed: the smallest one found in the cell library and the biggest one (20 times the size of the former).

The simulations showed that the inverters present high delay variability, closely modeled by a Gaussian curve. It can be described by the mean of the results, and the relative standard deviation. Figures 4.2, 4.3 and 4.4 shows the results of rising transitions delay, for all combinations of inverters and sources of variability (global, local and both). The exact mean values had to be omitted due to confidentiality reasons, but they are in the range of a typical 90 nm process.

The results show that, for the small cell, the global variability is about two times larger than the local variability. However, in the larger cell, the difference between them is more than 10 times larger, due to the great decrease of the impact of the local variations.

G. Neuberger et al., *Protecting Chips Against Hold Time Violations Due to Variability*,
DOI 10.1007/978-94-007-2427-3_4,

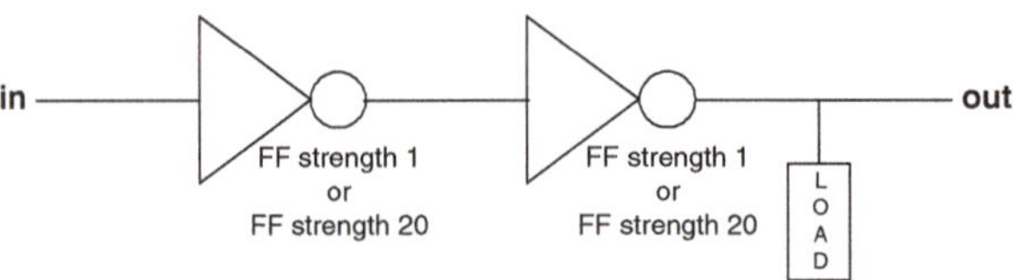

Fig. 4.1 Logic circuit to investigate delay variability

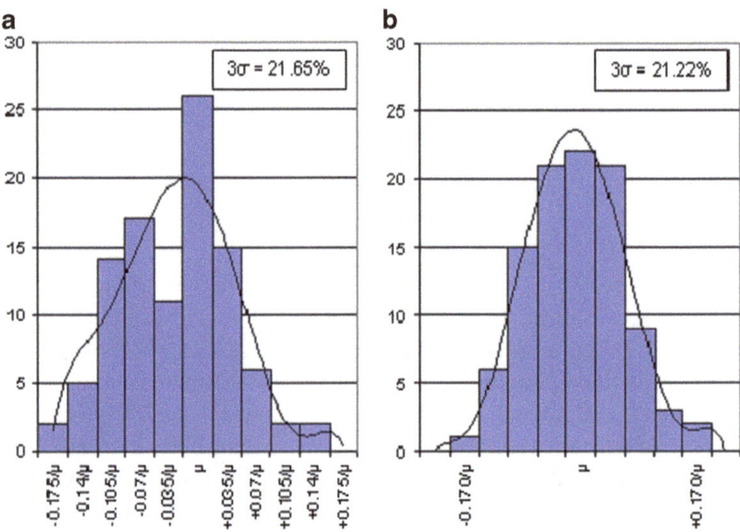

Fig. 4.2 Monte Carlo simulation results for rising transitions, only global variations: (**a**) small inverters (1×) and (**b**) large inverters (20×)

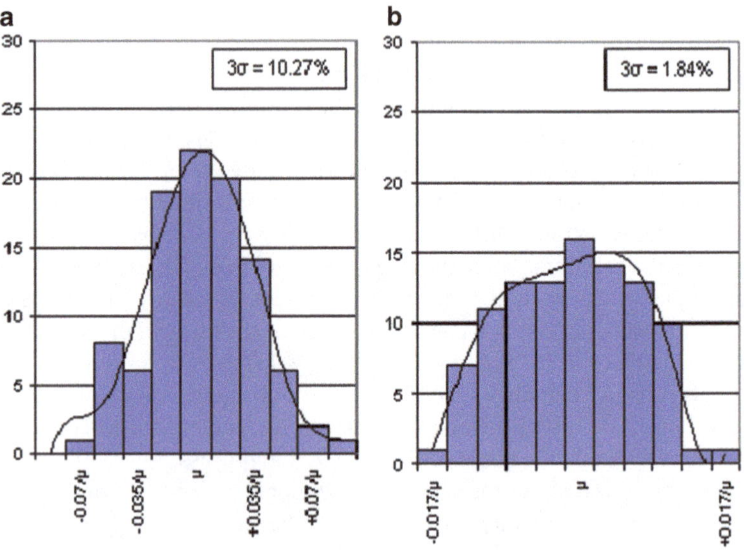

Fig. 4.3 Monte Carlo simulation results for rising transitions, only local variations: (**a**) small inverters (1×) and (**b**) large inverters (20×)

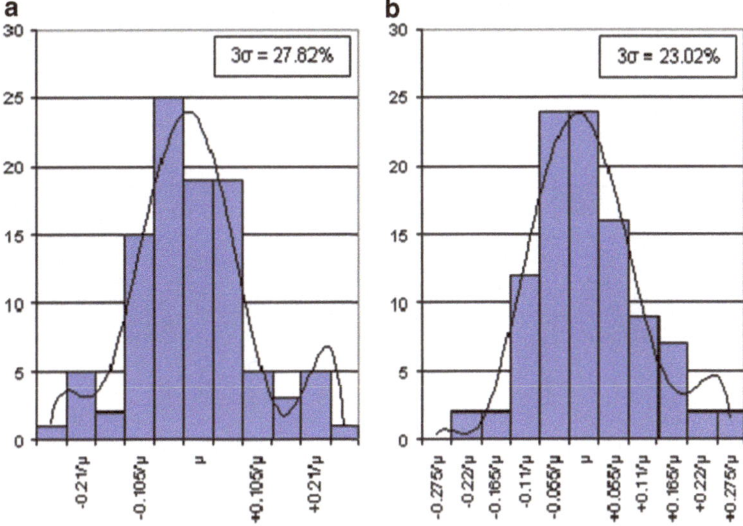

Fig. 4.4 Monte Carlo simulation results for rising transitions, both variations: (**a**) small inverters (1×) and (**b**) large inverters (20×)

4.2 Flip-Flop Sensitivity

The sequential elements must also be characterized in its sensitivity to variations. Three different parameters are important in this case: delay, setup time and hold time. The FFs to be tested are master–slave edge-triggered type D FFs. They are of the smallest driving capacity in the library. Besides the D and CLK inputs, the FF also has the signals RN, TI, TE. RN is a reset signal, while TI (test input) and TE (test enable) are for test purposes, to be used in a boundary scan chain. These inputs help to decrease the hold time, and they were used instead of normal FFs because they are commonly used in commercial products to increase testability.

The same methodology was applied, using Monte Carlo simulations to measure the mean and the deviation of the delay, setup time and hold time. Figure 4.5 shows the results for rising transitions.

The results show that the FF delay has a smaller relative variability than the inverters, but the absolute variability is about three times larger, because of the much larger FF delay. The setup and hold time deviation have about the same value as the inverters (because they are not as large as the delay).

4.3 Hold Time Violation Probability

Examining the sensitivity of FFs and inverters, one can notice that they can vary by several picoseconds. Now let's put it all together to look into the probability of a hold time violation being produced. Figure 4.6 shows the circuit initially used to measure this.

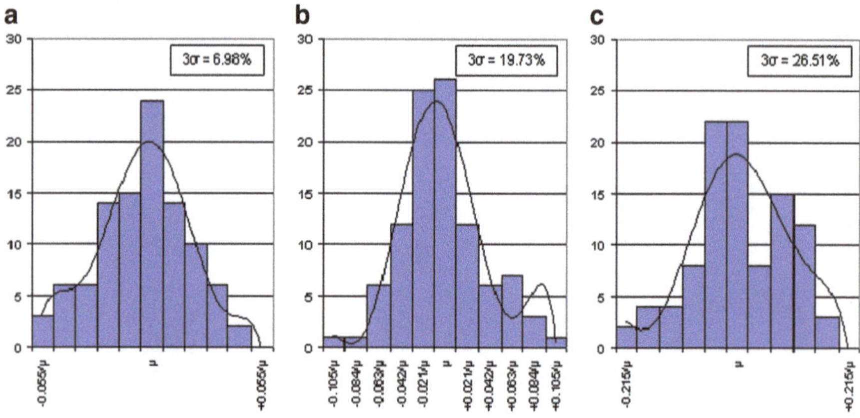

Fig. 4.5 FF parameter variability: (**a**) delay, (**b**) setup time and (**c**) hold time

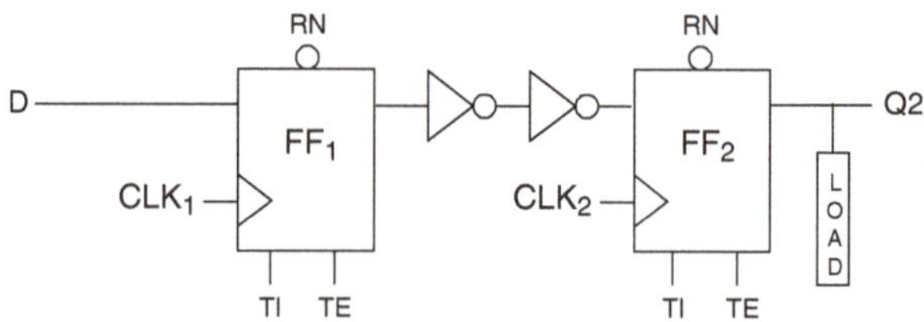

Fig. 4.6 Circuit with FFs and inverters to check hold time violation

It has a logic path between two FFs, and the FFs have two different clock inputs. By delaying CLK2 in relation to CLK1, we are producing an artificial clock skew.

To detect a violation, the FFs must be initialized with opposite values, and then a transition must be applied in the input. If there is no clock skew, FF1 will change to the new input value, and FF2 will change to the previous FF1 value. If the clock skew is increased, the circuit still operates correctly while the skew is smaller than the delay of the logic path. Using a parametric simulation, the skew was increased by steps of 1ps, until the FF2 does not sample the correct value, characterizing a hold time violation. The exact skew which makes the path fails is called critical clock skew, and is determined by a parametric simulation using the nominal process parameters.

Using this methodology, the critical clock skew was found for this configuration and also for other ones. It was compared with the typical clock skews achieved by complex circuits in the same technology. They were in the same magnitude order, showing that it is a potential problem.

The next step was to determine the impact of process variations on the hold time violation probability. To do so, the skew was fixed as the critical clock skew, and Monte Carlo simulation was performed. Then we can see the percentage of times that the circuit will fail, even though it should work using nominal parameters. The percentage of violations in this case was 48 %. To have an idea about the standard deviation, the simulation was repeated decreasing and increasing the skew by 3 ps. The percentage of failing circuits was 17 % and 82 %, respectively. This showed that a fine resolution is needed for the precise measurement of the critical skews, since within a small range of 3 ps, the difference between the percentage of violations was large.

4.4 Test Circuits for Fabrication

After examining that the hold time violations are a real possibility, we must decide which circuits will be fabricated in silicon to provide a wide analysis of the variability. Four different paths were then chosen.

The two basic configurations are two simple pipeline stages with two master–slave edge-triggered FFs without logic between them, representing one stage of a scan chain. Further pipelines including six small inverters between the FFs, represent short logic paths. The FFs used in this work are conventional rising edge-triggered master–slave FFs composed of CMOS transmission gates in the forward propagation path and C^2MOS latches in the feedback loops [19] with typical library extensions such as input and output node isolations and local clock buffers.

For each configuration a version with the weakest FF of the standard cell library, i.e. smallest transistor sizes and hence largest sensitivity to process variations, and a version with 8x increased driving strength is used. Comparing the results of both it is possible to analyze the impact of different transistor dimensions on the variability. The inverters used in both versions are of the minimum size, since these configurations represent typical non-critical paths where large driving capability is not required. Figure 4.7 shows the different four combinations of circuits under test.

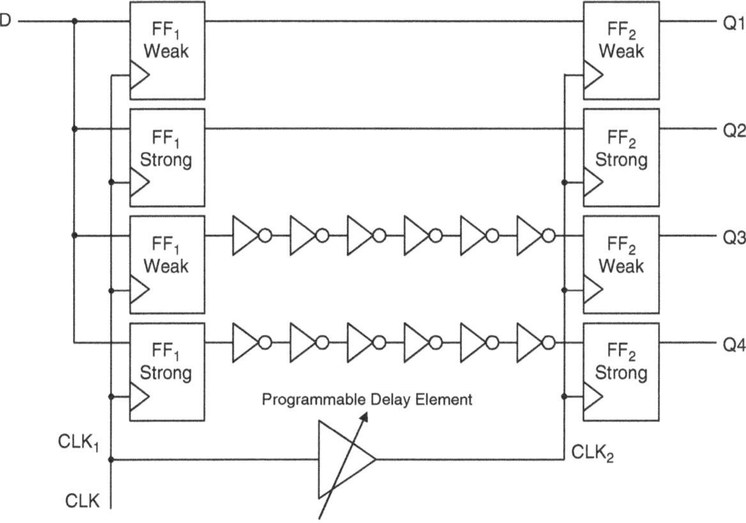

Fig. 4.7 Circuit with FFs and inverters to check hold time violation

To emulate clock uncertainties, the sending and receiving FFs are controlled by different clock signals. A programmable delay line generates the clock signal CLK2 of the receiving FFs. If this artificial clock skew is large enough, i.e. CLK2 arrives after CLK1 and exceeds the internal race immunity $t_{CLK\text{-}Q}$-t_{HOLD} of the FF, a race is produced and detected if the output of both FFs are of same value at same time ($Q1(t)=Q2(t)$). The violation can be detected by initializing the FFs with opposite values, and applying a pulse in the data input. As long as $Q1(t) \neq Q2(t)$ pipeline operation is correct.

The programmable delay element must be able to provide an adequate resolution for good statistics. By simulation, it was verified that a resolution that would be able to achieve in the available technology was about 1ps. The next section shows the development of the circuitry able to test the circuits showed in this section.

Chapter 5
Measurement Circuits

For Precise Characterization

In this chapter, it is shown all circuits that are needed to perform the measurement of the test circuits on wafer, as published in [33]. First, the delay line that will produce the artificial skew between the FF clocks is discussed. Then the ring oscillator needed to calibrate the delay line is showed. The next step is the design of a shift register to reduce the number of inputs. Finally, we put it all together and show the final layout.

5.1 Programmable Delay Line

As stated in the previous section, we need a scheme to produce an artificial skew between the clocks of the two FFs in the circuit under test. It must be programmable over different skews, and must have an accurate enough resolution for the given technology, to produce good statistics. The chosen structure able to do this was a programmable delay line.

To specify the critical clock skew producing a hold time violation, the artificial skew is programmable over a wide range of 80 steps corresponding to a resolution of ~1 ps. The delay line is composed of two inverters, and 80 NMOS gate capacitances as load elements connected to the inverters via pass transistors. The capacitances and transistors were carefully designed to be able to achieve steps of the desired resolution.

For coarse-grain clock skew shifting a multiplexer to enable or disable a further buffer chain is added. It is needed because the versions with 0 or 6 inverters have very different critical clock skews, so the buffer chain needs an equivalent delay of about 6 inverters. Figure 5.1 shows the implemented circuit.

An essential part of the delay line is the design of the load elements (the NMOS and PMOS pass transistors, and the NMOS gate parasitic capacitance). Different transistor widths and configurations were tried until the resolution of 1 ps could be achieved. Even the extracted parasitic capacitances in the layout had a great impact in the achievable delay, and caused a redesign in the layout.

G. Neuberger et al., *Protecting Chips Against Hold Time Violations Due to Variability*, DOI 10.1007/978-94-007-2427-3_5,

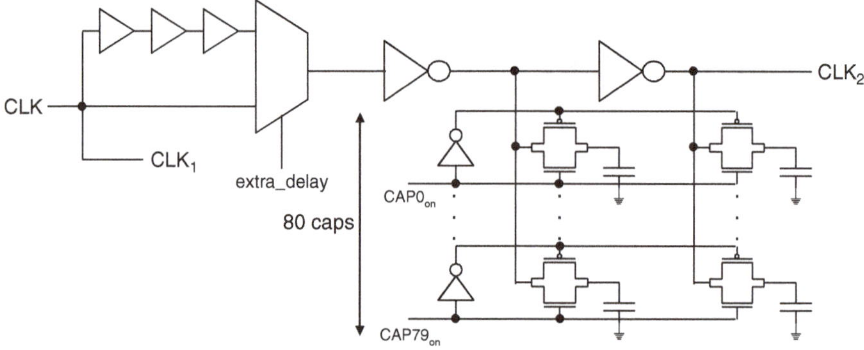

Fig. 5.1 Schematics of the programmable delay line for the clock skew emulation

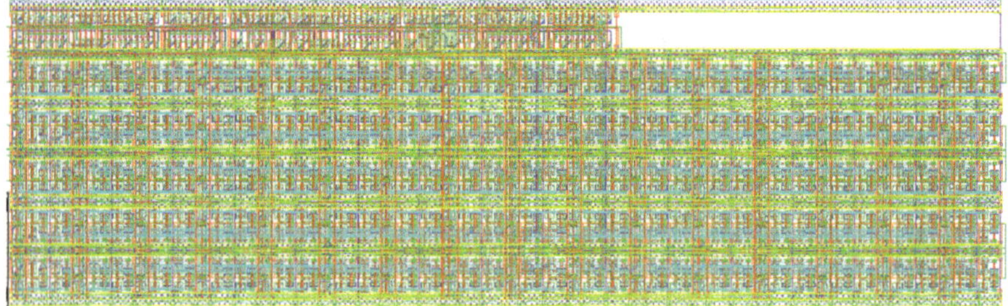

Fig. 5.2 Final layout of the programmable delay line

The final layout of the complete programmable delay line is shown in Fig. 5.2. The first row is composed by the two inverters, multiplexer and buffers, while in the other rows there are the 80 programmable parasitic capacitances.

5.2 Ring Oscillator

Once the delay line is designed, we must use a method to measure the delay that is actually programmed on it. It is not possible to just program a specific number of capacitances and consider that the delay is the one achieved by simulation. It is necessary to measure the delay externally.

To measure the absolute time produced by a specific setting of the programmable delay line, it is additionally placed in the middle of a ring oscillator. The ring oscillator is connected to an 11-stage frequency divider to monitor the output frequency. Thus, it is possible to determine the programmed delay based on measuring and comparing the frequencies achieved with different numbers of capacitances.

The ring oscillator was designed with 17 inverters plus the delay line. In the output of the oscillator, 11 stages of a frequency divider were used to achieve externally a frequency that the available equipments would be able to measure with

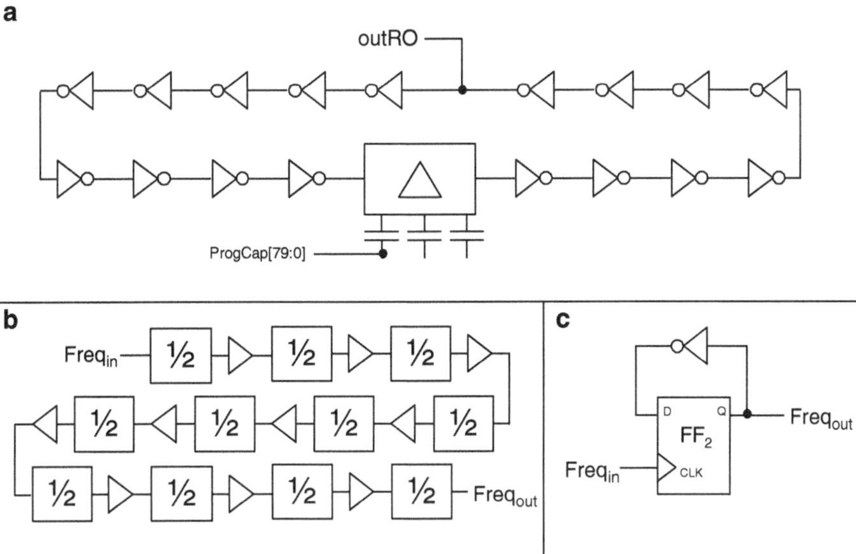

Fig. 5.3 Schematics: (**a**) ring oscillator with delay line, (**b**) frequency divider and (**c**) one stage of the frequency divider

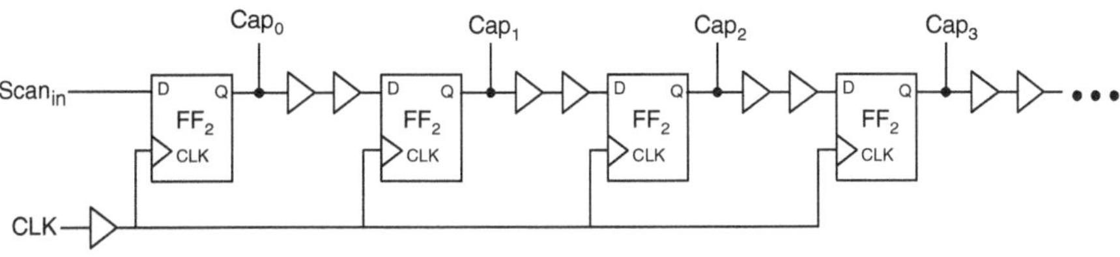

Fig. 5.4 Schematic of the shift register

enough precision. The frequency divider is very simple, composed by a FF and an inverter. Figure 5.3 shows the schematics of these components.

5.3 Shift Register

After the design of the delay line, it was verified that many input pins would be needed to program all over the 80 different configurations. To reduce the number of pins, the methodology of programming was changed to serial. A chain of 80 flip-flops in a shift register scheme was implemented to allow the delay line programming. By doing so, the 80 pins were substituted by only two: scan_in for data input and CLKscan to control the FFs clock. The drawback is the extra time needed for programming before the measurement itself. But it was the only option, since the available probe card had only 24 pins. Figure 5.4 shows the shift register. Two buffers were included between each FF to avoid hold time violations.

5.4 Final Circuit and Layout

After the design of all blocks, it is time to put them all together. Figure 5.5 shows the final block-based schematic. In the top right are the four circuits under test. They are driven by the same input, but they have each one its own output. The CLK1 signal is provided externally for the test circuits and the delay line. The delay line then creates the delayed CLK2 and sends it to the four paths. Both the delay line that sends the CLK signals and the one in the middle of the ring oscillator are programmed by the shift register. The shift register needs one data and one clock input. The output of the ring oscillator goes to the 11-stage frequency divider. The ring oscillator and the frequency divider use a different Vdd than the rest of the circuit, to avoid voltage drop due to the high switching activity. All the inputs and outputs of the circuit are buffered.

Looking to the circuits, it is possible to see that the total number of PADs needed is 15. This is less than the number of PADs in the available probe card (24 PADs). The circuit was then manufactured and it was be possible to measure it without problems. The final modification must include a buffer able to drive the PADs. To design it, we must use a PAD model. The buffer able to drive it was composed of a set of buffers in increasing scale. Figure 5.6 shows the PAD model and the driving buffer.

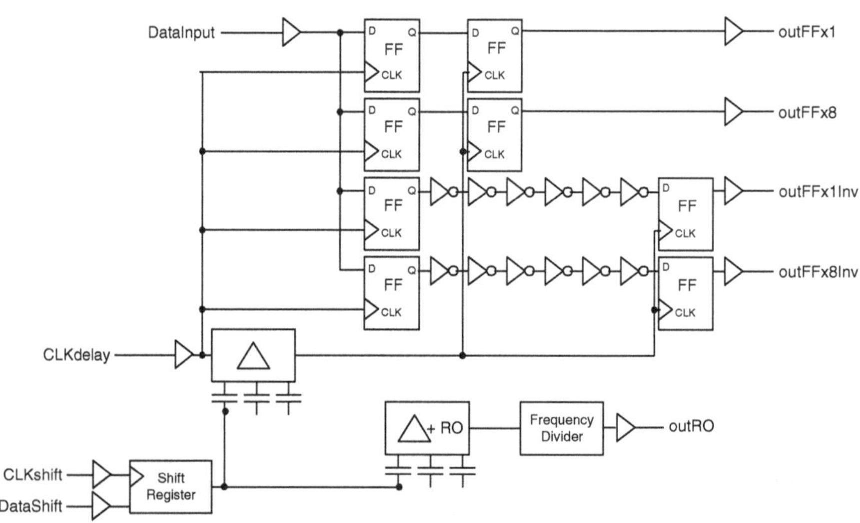

Fig. 5.5 Final schematic of the fabricated circuit

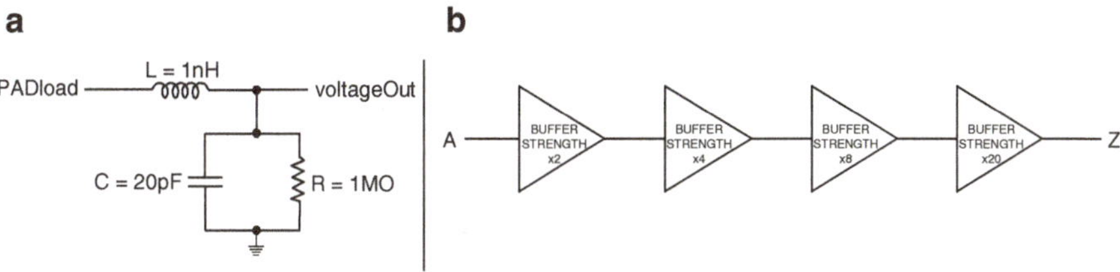

Fig. 5.6 (**a**) PAD model and (**b**) driving buffer

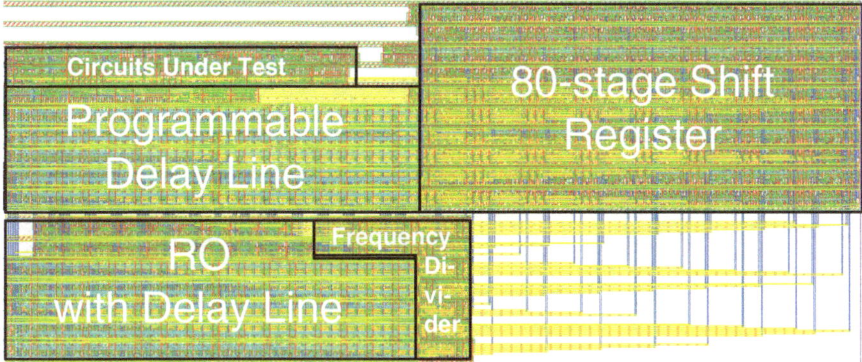

Fig. 5.7 Final 130 nm layout of the complete fabricated circuit

After the design of the circuits, the layout is straightforward. The circuits are very regular, resulting in simple placement and routing. A full-custom methodology was used, to minimize area and delay and to decrease the impact of process variations in parts other than the circuits under test. Figure 5.7 shows the final layout in a 130 nm technology and its blocks, while in Fig. 5.8 it is possible to see the 90 nm layout.

The final step is to route the inputs and outputs to the PADs. A template for the PADs, with 24 of them, was used, but not all of them are needed. The input PADs were placed in the left, the outputs in the right, and power (the two pairs of Vdd and Gnd) in the middle. It is shown in the Fig. 5.9.

After the complete layout was finished, a final design rule check was performed, and simulations with extracted parasitics. It worked as expected. The circuit is ready for fabrication.

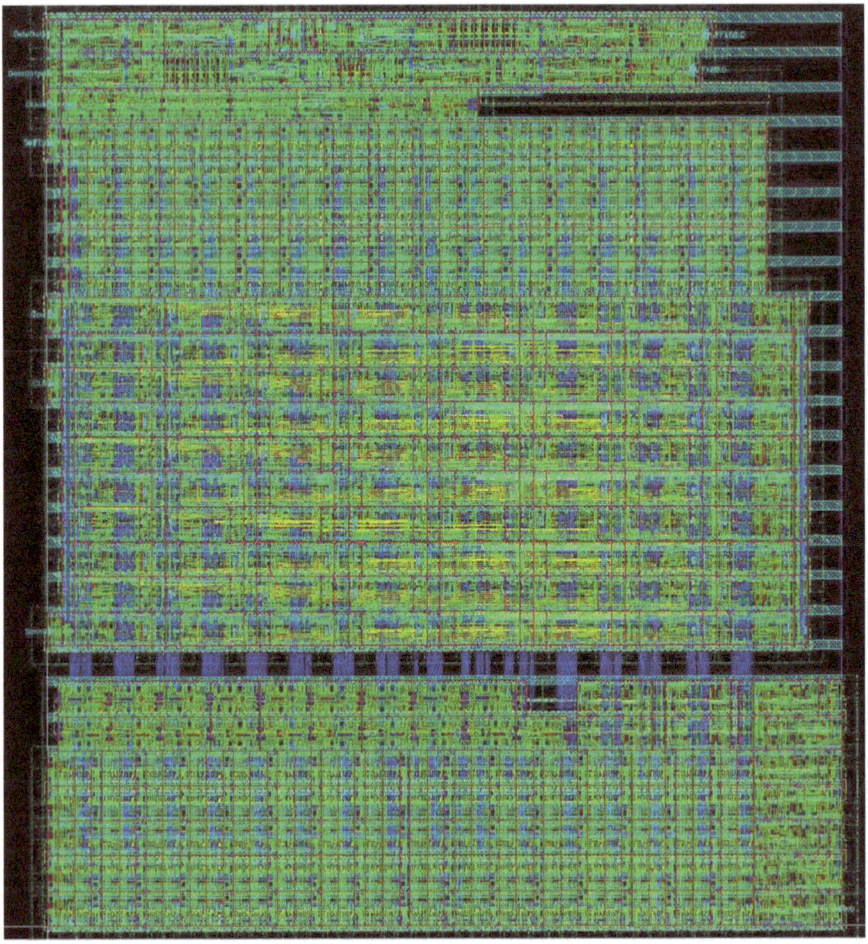

Fig. 5.8 Final 90 nm layout of the complete fabricated circuit

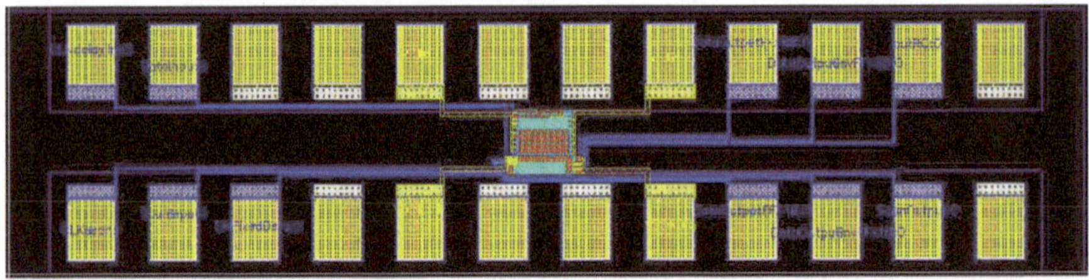

Fig. 5.9 Final layout with PADs

Chapter 6
Experimental Results
Of Fabricated Circuits

In this chapter, we discuss the setup of the equipments needed to perform the measurements, and results found in these measurements. First, the basic setup is shown; and then the measurement flow followed in each circuit measurement is drawn. All the measured data with the different combinations (technologies, temperature, Vdd's) is shown in the second part of this chapter.

6.1 Measurement Flow and Setup

The circuits were fabricated in 130 and 90 nm low power CMOS technology, and two wafers in 130 nm and one wafer in 90 nm were measured. Each 130 nm wafers has 182 chips, but the 90 nm wafer only 36 (there were other circuits on the same die, leading to a larger die size). The total size of the 130 nm chip is 180×71 µm, while the size of the 90 nm is 70×77 µm. The measurements are done at different temperature and voltage source conditions.

The wafers were not sliced and encapsulated for individual measurement, but measured directly with an Süss prober, to allow a spatial analysis of the results over the wafer. The prober has its temperature controlled.

A Tektronix DG2030 data generator was used to provide the inputs. It has eight independent sources, but only six of them were used (data input and clock input for the test circuits, data input and clock input for the shift register, a signal to choose between the two delay line configurations, and a signal to trigger the oscilloscope). The oscilloscope was a Tektronix TDS5054. It has only four channels, while five are needed (the four test circuit outputs, and the ring oscillator output). So, the measurement had to be done in two steps, changing the cables between them: first the measurement of only the ring oscillator frequencies, and second the test of the circuits. Moreover, two voltage sources were used to provide the power to the circuit.

All the equipments were connected to a PC, and the measurement process was partially automated via the PC using the software LabVIEW. The LabVIEW programs were carefully designed and modularized, so there are different modules

G. Neuberger et al., *Protecting Chips Against Hold Time Violations Due to Variability*,
DOI 10.1007/978-94-007-2427-3_6,

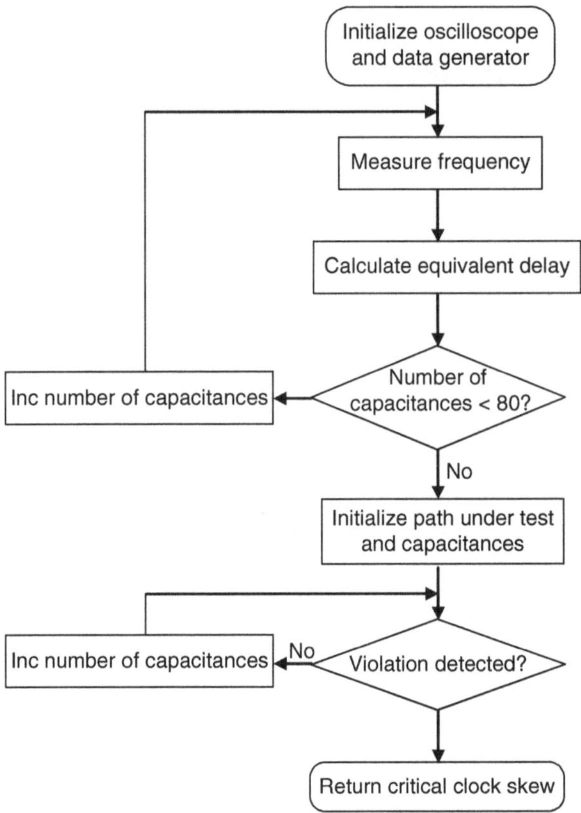

Fig. 6.1 Measurement flow

for all different tasks (data generator initialization, oscilloscope initialization, frequency measurement, etc). In the end, there is one top-level module that uses them all to perform the complete measurement.

For the measurement, first the settings for all combinations of the 80 capacitances are written into the shift register. Then the frequencies of the ring oscillator are measured for all configurations to calibrate the programmable skews. For measurement of the delay variations of the logic path, the delay line is initialized with minimum delay, and the delay is stepwise increased until a violation in the pipeline is detected. The corresponding delay estimated from the ring oscillator measurements is the critical clock skew for the given die and operating conditions. The procedure is repeated for each of the four test circuits considering the rising and falling input transitions. Figure 6.1 shows the measurement flow.

6.2 Ring Oscillator Frequency Variability

The first experiment was to measure and compare the maximum frequency that the ring oscillators can achieve (when 0 of the capacitances are turned on). In the 130 nm wafers, it shows a typical global wafer variation with slower dies in the center of the wafer. In the 90 nm wafer, the distribution is more random. The frequencies are normalized to a given value, to omit derivation in the absolute speed values of the technological data. Figures 6.2 and 6.3 show the 130 nm frequency variability in both measured wafers, while the 90 nm results are in Fig. 6.4. The figures show the measurements using the following conditions: temperature = 25° and nominal Vdd (1.5 V in 130 nm, 1.32 V in 90 nm). Additional temperature and Vdd conditions were also used, but they will not be shown here, since they are very similar, when using a normalized axis.

The frequencies in 130 nm wafer 1 are larger than in wafer 2, and also its relative variability.

The next step was to measure the frequencies achieved by programming more capacitances, and to calculate the resolution provided by the programmable delay line. The faster circuits achieve resolutions less than 1 ps, while none of the chips had a resolution of more than 1.2 ps.

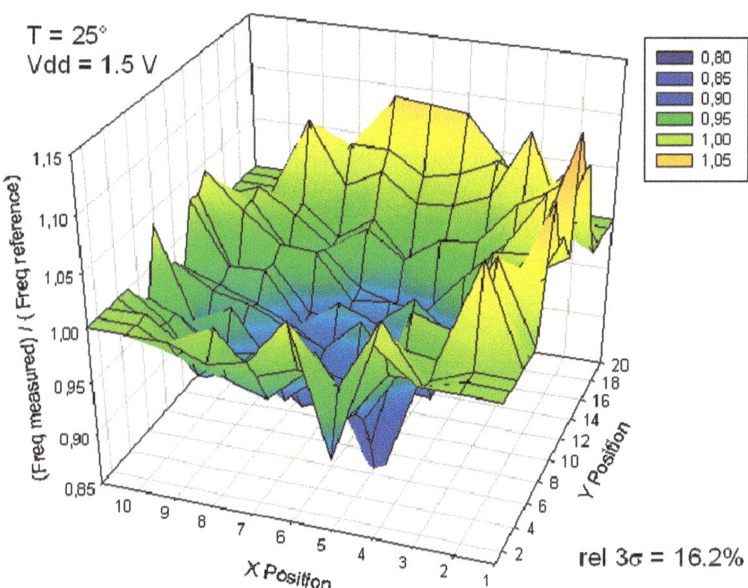

Fig. 6.2 Normalized frequency variability in 130 nm technology over wafer 1

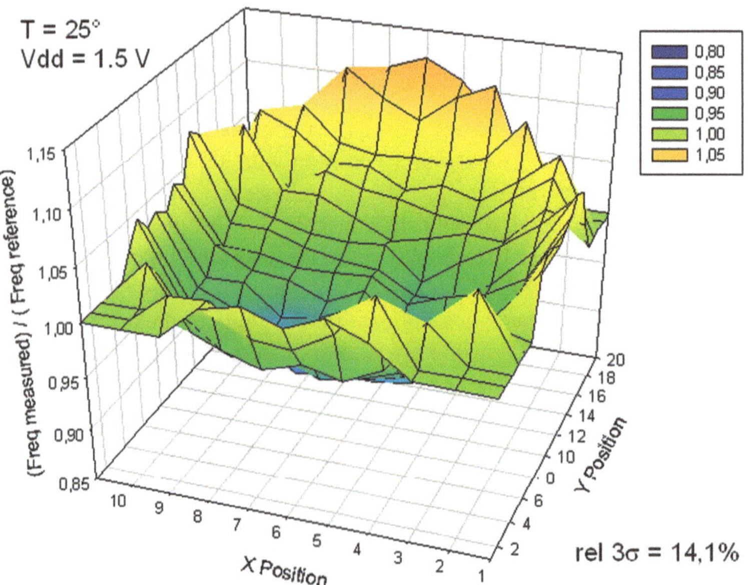

Fig. 6.3 Normalized frequency variability in 130 nm technology over wafer 2

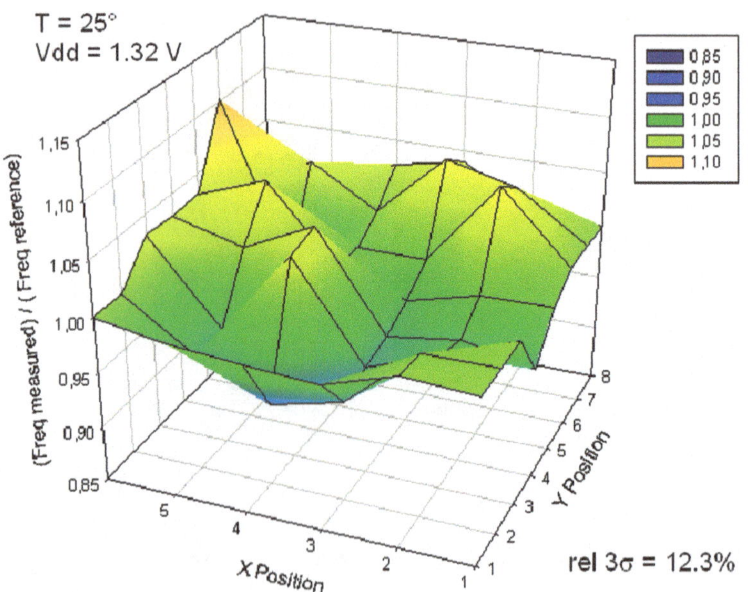

Fig. 6.4 Normalized frequency variability in 90 nm technology over the wafer

6.3 Hold Time Violation Distribution

After calibrating the delays that can be programmed by each step of the delay line, it is time to measure when the test circuits fail and to save its critical skew. The process was repeated for all dies in all wafers.

Figures 6.5 and 6.6 shows the distribution of the critical clock skew for 0-1 transitions in all 4 test circuits for the two 130 nm wafers, and Fig. 6.7 for the 90 nm wafer. The nominal case (mean critical skew) is set to 0ps. The expected Gaussian curve for normal distributions is observed. Based on this data and repeating the measurement procedure for 1-0 transitions, the mean critical clock skew and the standard deviation are extracted. Tables 6.1, 6.2 and 6.3 summarize these results. The results are normalized to the first test circuit of each technology, so it is not possible to compare between the results of different wafers, but it will be done in the next subsection. These tests were repeated for different measurement conditions, like temperature and Vdd.

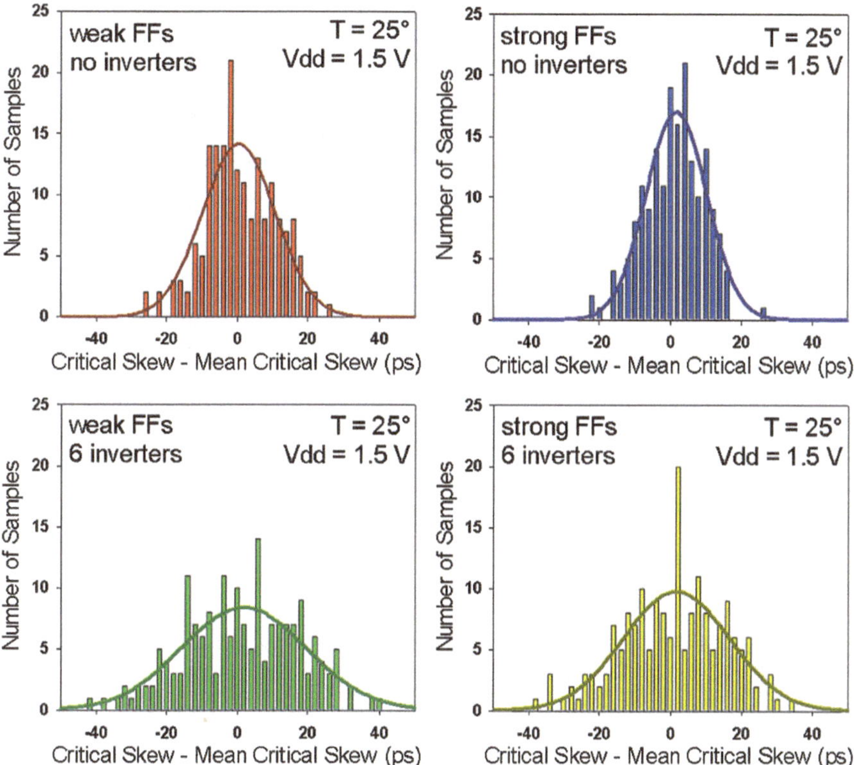

Fig. 6.5 Measured distribution of the critical clock skews for rising transitions in 130 nm technology wafer 1

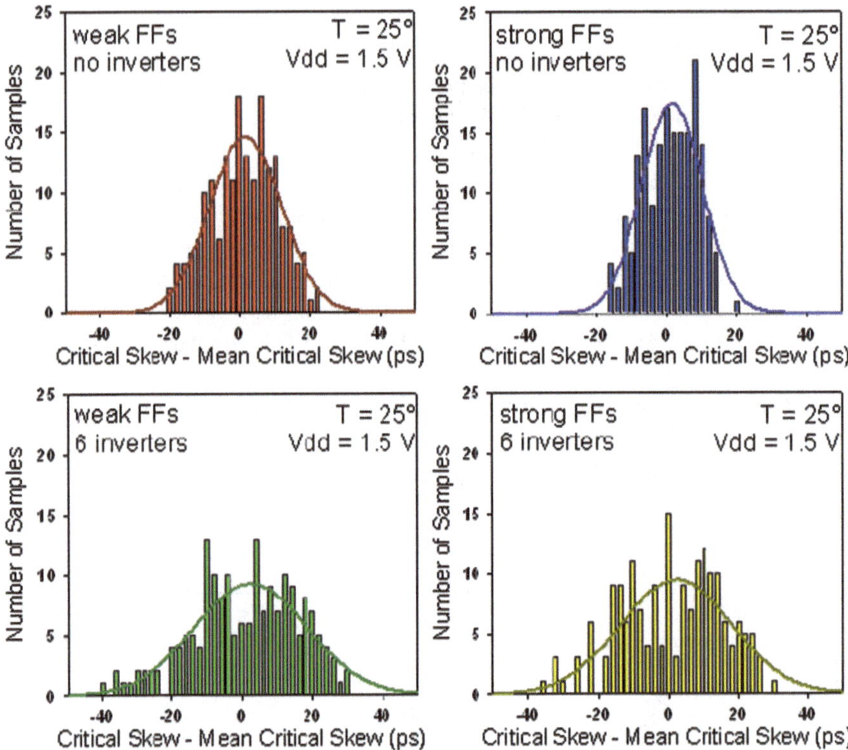

Fig. 6.6 Measured distribution of the critical clock skews for rising transitions in 130 nm technology wafer 2

The 3σ deviation of the delay can be up to 19 % of the nominal value. The critical skews are in the range of the clock skew that can be expected in circuits using the same technology, showing that these statistical effects have to be considered during hold-time fixing at the end of the layout generation. It is important to note that using larger FFs, the absolute variation of the critical skew decreases, but the relative value remains similar, since these circuits are faster. This indicates that larger FFs have an increased probability of violation, since the clock skew needed to provoke the failure is smaller.

The test circuits with FFs and extra inverters have presented larger absolute variability than the version without inverters, as expected because the effects of the FFs and inverters variability are combined. However, it has shown a smaller relative variability, which brings the following conclusions. One possibility is that the FFs are more sensitive to process variations than the inverters, so a circuit with the two components in average has a total smaller deviation. Another explanation is that a large number of inverters in the path can compensate the average local variability.

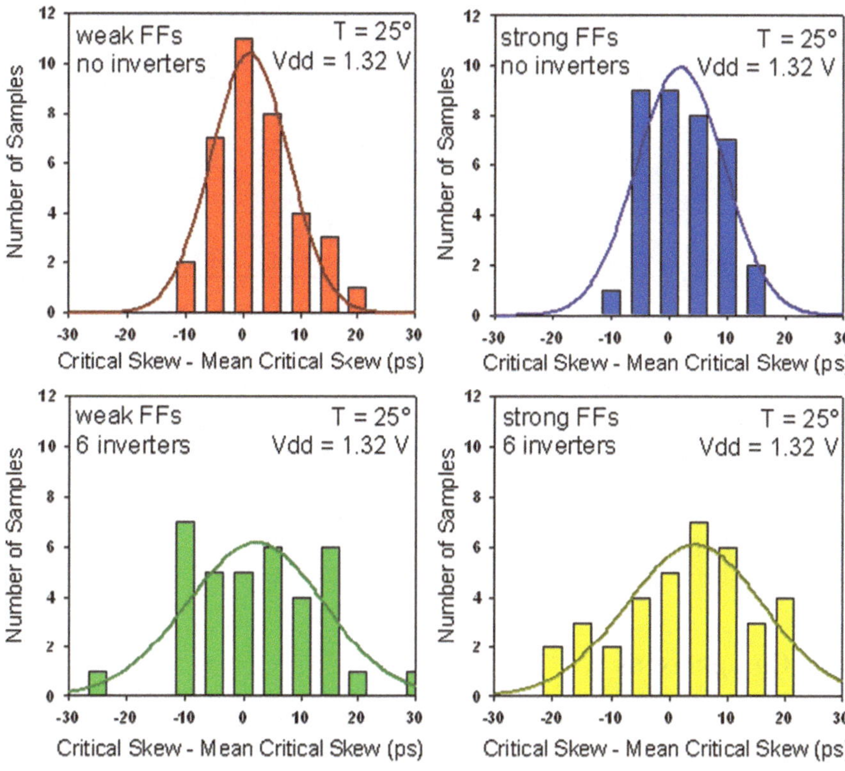

Fig. 6.7 Measured distribution of the critical clock skews for rising transitions in 90 nm technology

Table 6.1 Normalized hold time violations in 130 nm wafer 1 at V_{DD}=1.5 V and T=25°

Circuit	Transition	μ	3σ	3σ/μ (%)
Weak FFs,	Rising	100.00	18.60	18.60
No Inverters	Falling	109.52	19.08	17.43
Strong FFs,	Rising	88.09	15.60	17.73
No Inverters	Falling	94.68	16.62	17.55
Weak FFs,	Rising	184.11	30.75	16.71
6 Inverters	Falling	194.81	30.24	15.54
Strong FFs,	Rising	172.93	27.57	15.96
6 Inverters	Falling	180.47	28.83	15.99

Table 6.2 Normalized hold time violations in 130 nm wafer 2 at V_{DD}=1.5 V and T=25°

Circuit	Transition	μ	3σ	3σ/μ (%)
Weak FFs,	Rising	100.00	14.88	14.88
No Inverters	Falling	109.95	14.60	13.28
Strong FFs,	Rising	88.87	12.06	13.57
No Inverters	Falling	95.00	12.28	12.93
Weak FFs,	Rising	181.70	23.76	13.08
6 Inverters	Falling	192.71	23.19	12.04
Strong FFs,	Rising	170.54	22.44	13.16
6 Inverters	Falling	177.52	22.22	12.51

Table 6.3 Normalized hold time violations in 90 nm wafer at $V_{DD} = 1.32$ V and T = 25°

Circuit	Transition	μ	3σ	3σ/μ (%)
Weak FFs,	Rising	100.00	16.49	16.49
No Inverters	Falling	118.29	17.98	15.20
Strong FFs,	Rising	93.83	15.05	16.04
No Inverters	Falling	116.12	17.10	14.73
Weak FFs,	Rising	177.82	26.14	14.70
6 Inverters	Falling	197.20	28.01	14.20
Strong FFs,	Rising	173.68	25.75	14.83
6 Inverters	Falling	197.99	27.07	13.67

Another important point is that the master–slave FFs used in the experiment typically have a small or even negative hold time, and consequently a high race immunity. Repeating the experiments for faster FFs with larger hold times that are used in high-speed designs, the results would be even more critical because in this case the race immunity is worse and consequently the clock skew needed to provoke a failure or violation is smaller.

6.4 Comparisons

The basic results found using the nominal test conditions were shown in the previous subsection. The experiments were repeated for different conditions, and now the results will be compared between them. The comparisons are between different wafers in the same technology, different technologies, different temperatures, and different Vdd's.

6.4.1 Different Wafers of a Same Technology

Two wafers were measured in 130 nm technology, and they did not show exactly the same results. The first wafer measured was about 20 % faster than the second wafer. Probably this has an impact in the critical clock skew to produce the hold time violation (race immunity), and in the relative standard deviation of the variability. Figures 6.8 and 6.9 shows the comparison between these different wafers.

As expected, the slower wafer has larger race immunity and average critical skews. And it has also smaller relative standard deviation (~4.5 % × ~6 %). With these results, it is possible to see that for the same circuits and same technology, the faster dies are more susceptible to hold time violations.

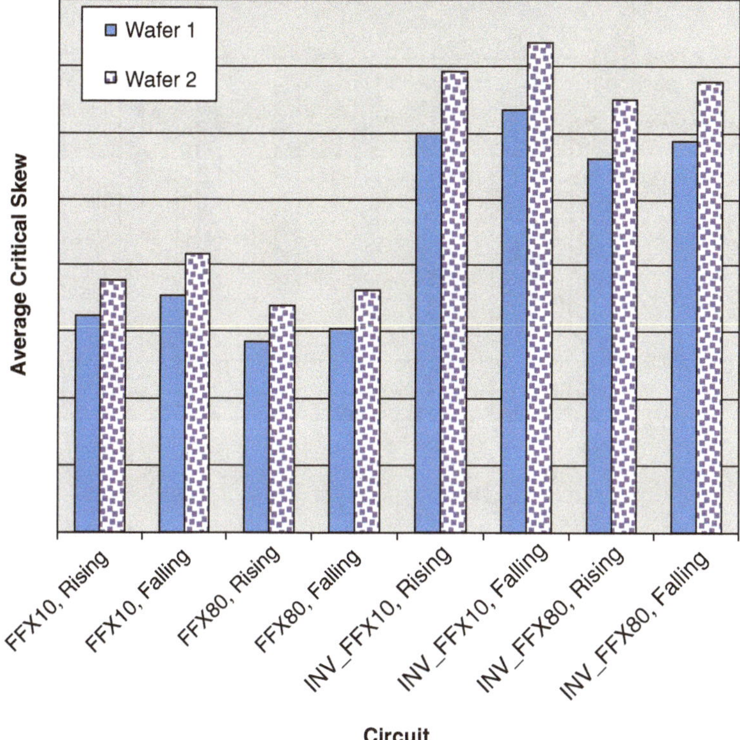

Fig. 6.8 Comparison between 130 nm wafer 1 and wafer 2 – average critical skew

6.4.2 Different Technologies at Nominal Conditions

Two different technologies were measured: 130 and 90 nm. The 90 nm technology is faster than 130 nm, and newer technologies are expected to be more susceptible to variations. However, the process control for this specific 90 nm technology is much better than for 130 nm, so the variations impact are not expected to explode at this node. Figures 6.10 and 6.11 show the comparison between 90 and 130 nm wafer 2.

The first graph shows that the internal race immunity is much smaller in 90 nm, as the FFs are faster. And the variability increases slightly (~4,5 % × ~5,5 %). So, 90 nm technologies are more sensitive to hold time violations, if the clock skew of the circuits does not scale in the same way.

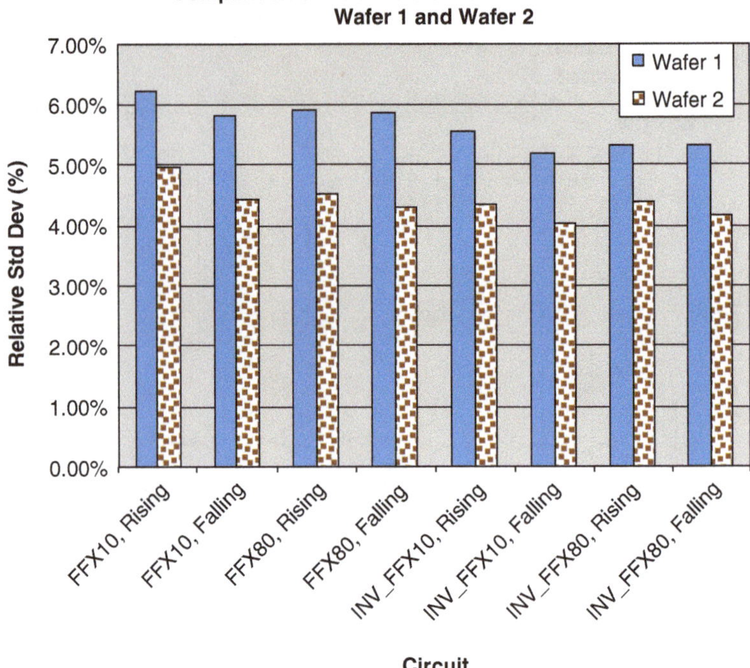

Fig. 6.9 Comparison between 130 nm wafer 1 and wafer 2 – relative standard deviation

Other important points are that the FF sizing does not change the internal race immunity in 90 nm, while in 130 nm the larger FFs have about 10 % smaller race immunity, probably due to minor changes in the schematic of the FF between the libraries of the different technologies.

6.4.3 Different Temperatures

It is important to also investigate the impact of temperature in the results. The experiments were performed in two different temperatures: 25 °C and 85 °C. But they are only for the 130 nm wafer 2, since the temperature did not show a big impact in the results and the measurements for the other wafers were not performed. Figures 6.12 and 6.13 show the comparison between 25 °C and 85 °C in 130 nm wafer 2.

The circuits operating at 85 °C are about 10 % slower than at 25 °C. As expected, the critical skew and race immunity are also 10 % larger. However, the relative variability remains about the same. As the clock skew is expected to scale in the same way [53], temperature does not have a large impact on hold time violation probability.

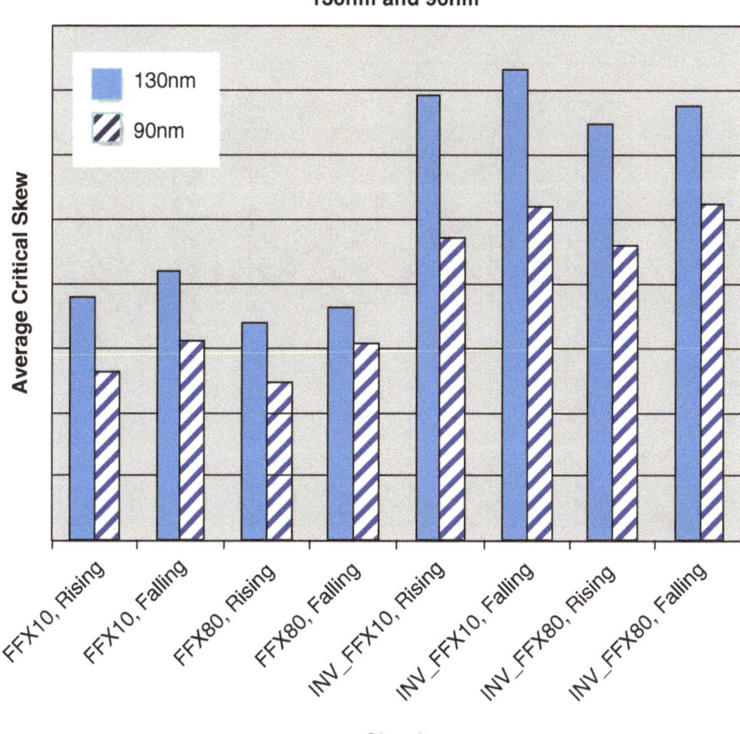

Fig. 6.10 Comparison between 130 wafer 2 and 90 nm wafer – average critical skew

6.4.4 Different Vdd's

Next, the influence of Vdd was investigated. The experiments were performed in three different conditions: nominal Vdd, 1.2, 0.9 V. The nominal Vdd is 1.5 V in 130 nm, and 1.32 V in 90 nm. The experimental results are for 130 nm wafer 2 and for the 90 nm wafer. Figures 6.14 and 6.15 shows the comparison when using different Vdd's in 130 nm wafer 2.

Decreasing Vdd, the circuits become extremely slower. The average critical skew scales in the same way (inversely proportional to Vdd). At 1.2 V, the circuits are about 50 % slower, while at 0.9 V, 180 % slower.

The relative standard deviation also increases, but not as strong. From ~4.5 % at 1.5 V, it goes to ~5.5 % at 1.2 V and ~7 % at 0.9 V. It is important to note also that more capacitances are needed to provoke the violation at smaller Vdd.

Figures 6.16 and 6.17 show the comparison when using different Vdd's in a 90 nm wafer.

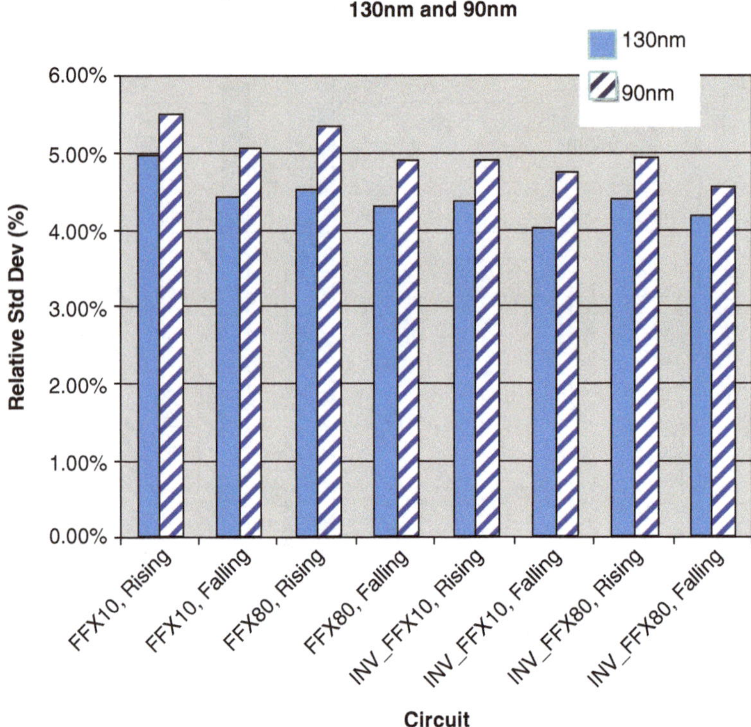

Fig. 6.11 Comparison between 130 wafer 2 and 90 nm wafer – relative standard deviation

The results in 90 nm follow the same results as in 130 nm, except that now the circuits are much faster. For example, the results in 90 nm 0.9 V are in the same scale as in 130 nm 1.2 V (0.9 V is a very small voltage for 130 nm, but not as much for 90 nm).

It is important to determine if the clock skew scales in the same way as the internal race immunity to check if the hold time violation probability changes. However, it is reported by the industry that decreasing Vdd can make some circuits having violations, to work again. These topics will be discussed in Chaps. 9 and 10.

6.4.5 Different Technologies at Same Vdd

Finally, a comparison with different technologies using the same Vdd instead of nominal Vdd was done. The experiments were performed in two different conditions: 1.2 and 0.9 V for both 130 and 90 nm. Figures 6.18 and 6.19 show the results for 1.2 V, while Figs. 6.20 and 6.21 show for 0.9 V.

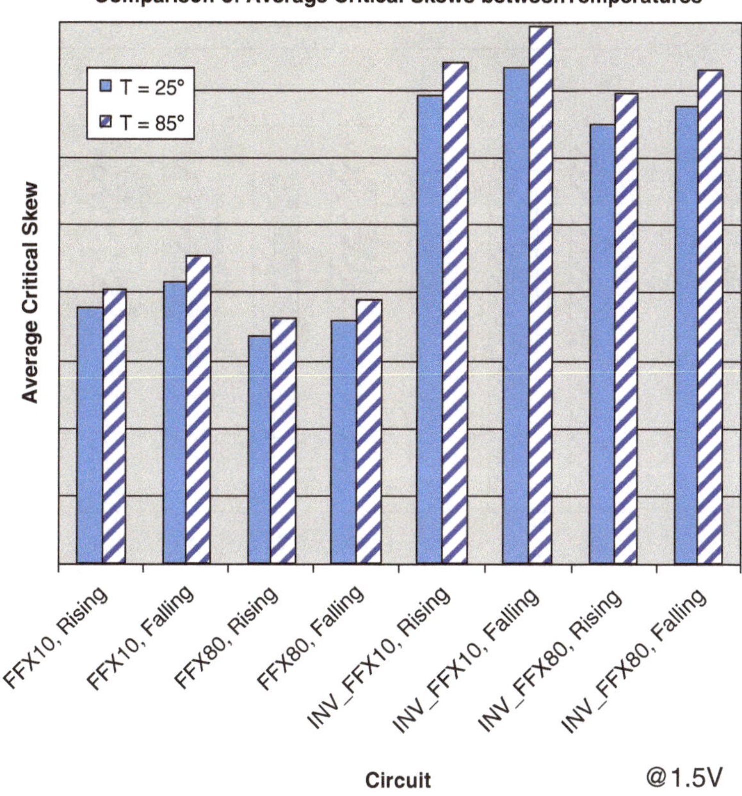

Fig. 6.12 Comparison between T=25° and T=85° in 130 nm wafer 2 – average critical skew

The results for 1.2 V show that in 90 nm, the race immunity is much smaller (almost half) than in 130 nm. However, the relative standard deviation increases, confirming that the variability is getting worse in newer technologies, especially if the clock skew does not scale in the same way. The results for Vdd=0.9 V confirms these points.

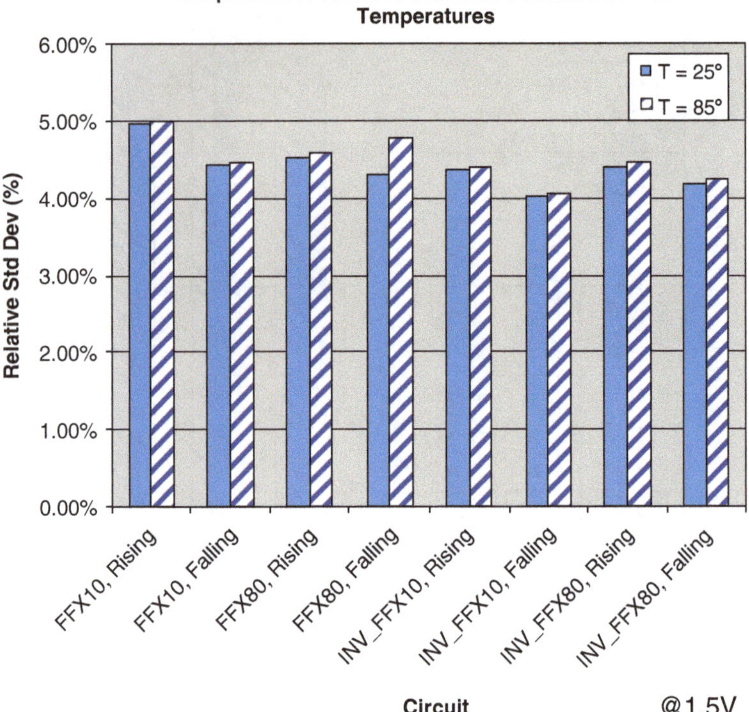

Fig. 6.13 Comparison between T=25° and T=85° in 130 nm wafer 2 – relative standard deviation

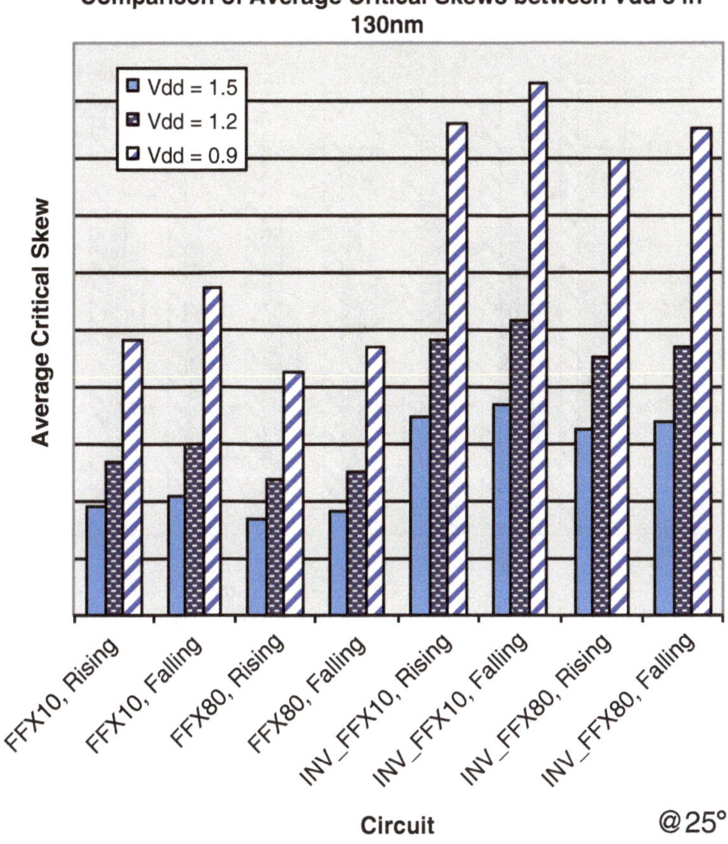

Fig. 6.14 Comparison between Vdd = 1.5, 1.2 and 0.9 V in 130 nm wafer 2 – average critical skew

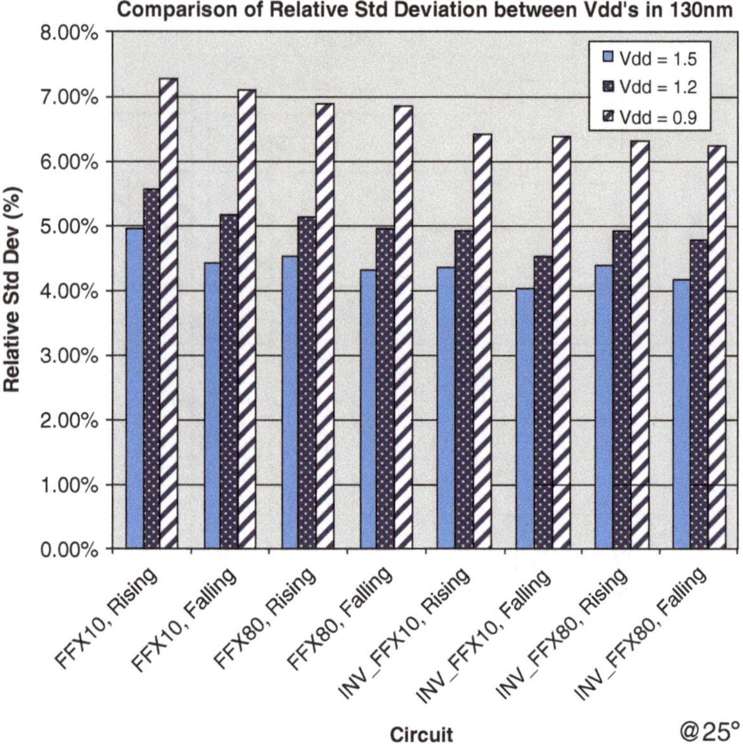

Fig. 6.15 Comparison between Vdd = 1.5, 1.2 and 0.9 V in 130 nm wafer 2 –relative standard deviation

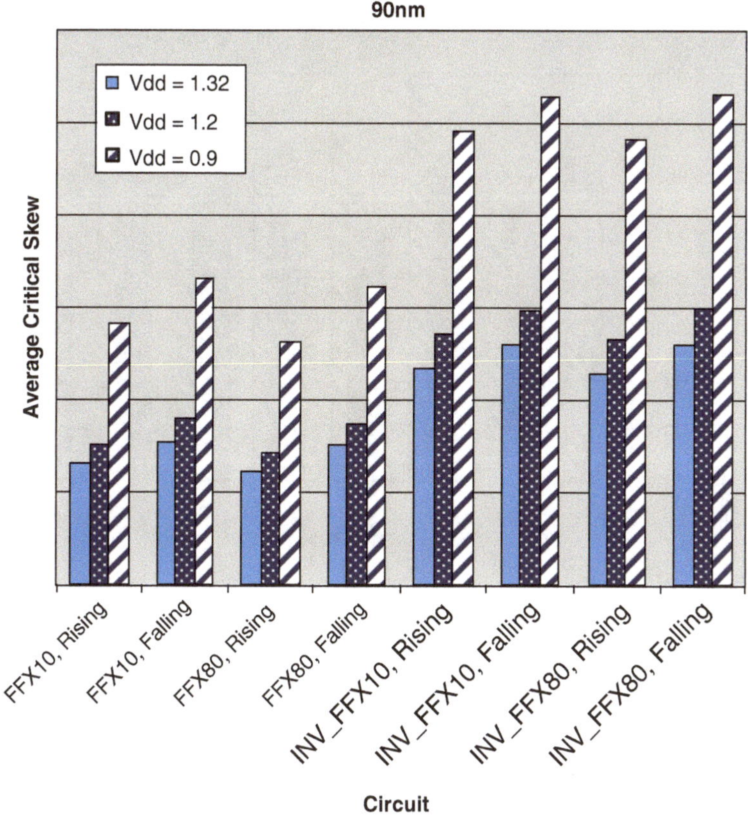

Fig. 6.16 Comparison between Vdd = 1.32, 1.2 and 0.9 V in 90 nm wafer – average critical skew

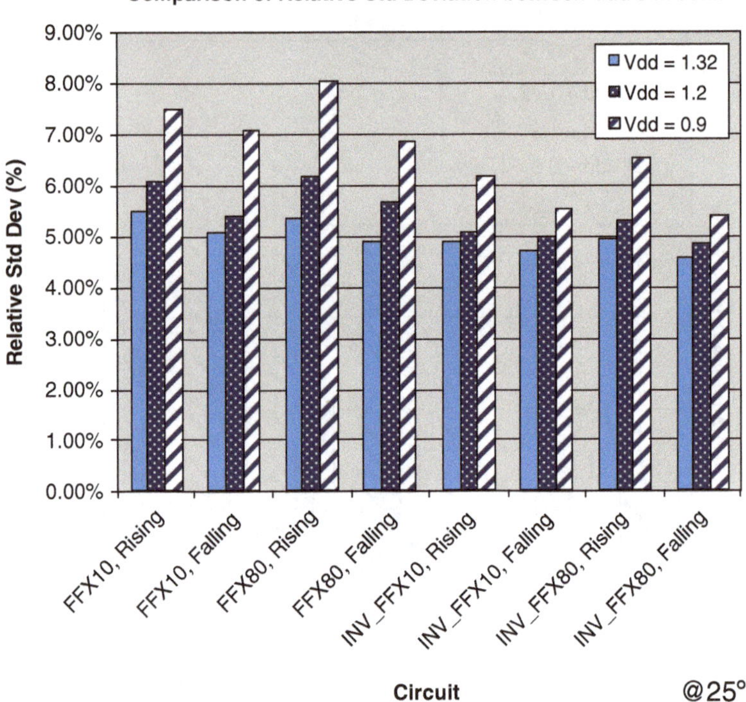

Fig. 6.17 Comparison between Vdd = 1.32, 1.2 and 0.9 V in 90 nm wafer –relative standard deviation

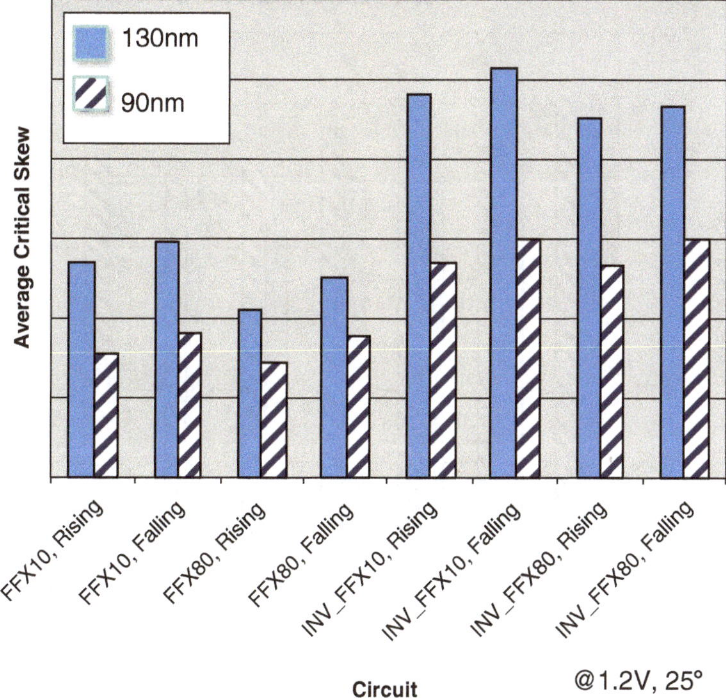

Fig. 6.18 Comparison between 130 wafer 2 and 90 nm wafer at Vdd = 1.2 V – average critical skew

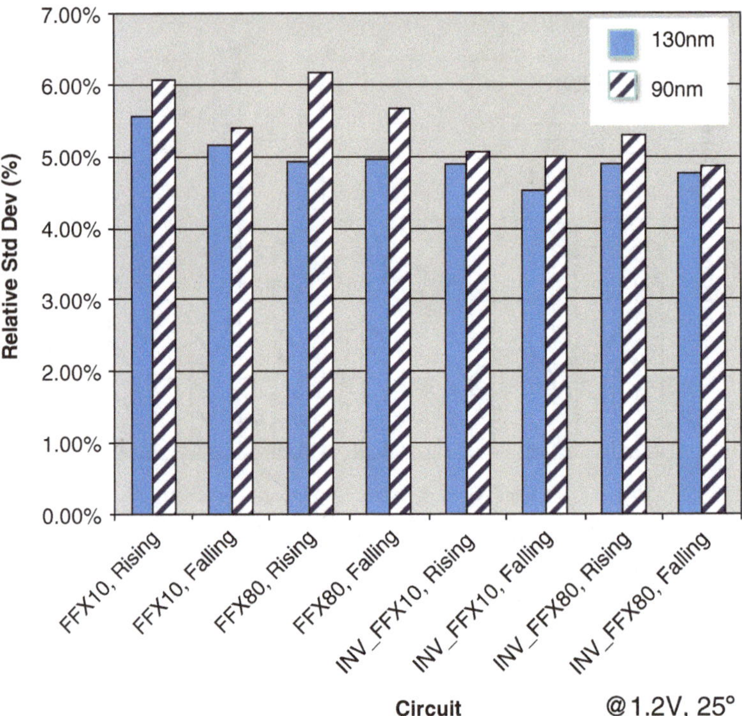

Fig. 6.19 Comparison between 130 wafer 2 and 90 nm wafer at Vdd = 1.2 V – relative standard deviation

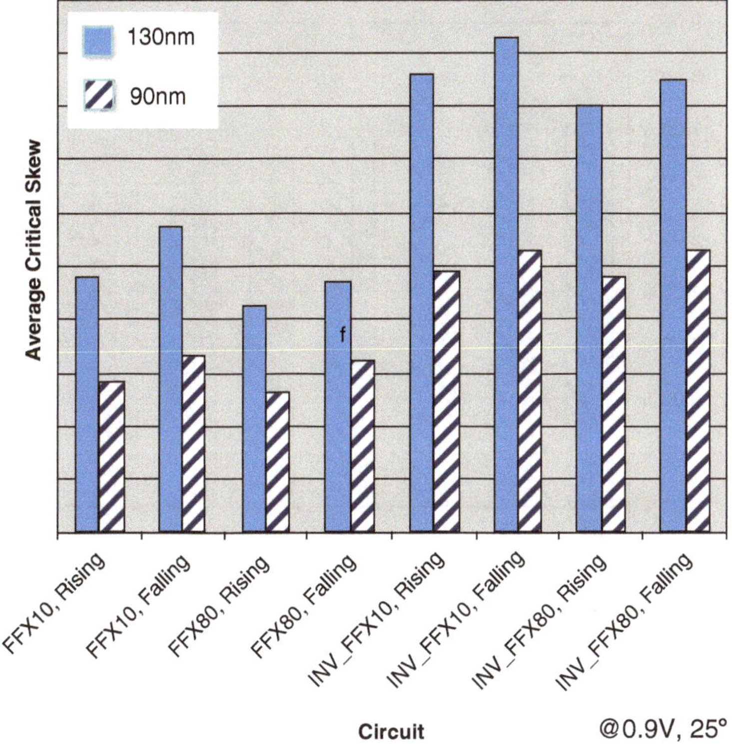

Fig. 6.20 Comparison between 130 nm wafer 2 and 90 nm wafer at Vdd = 0.9 V – average critical skew

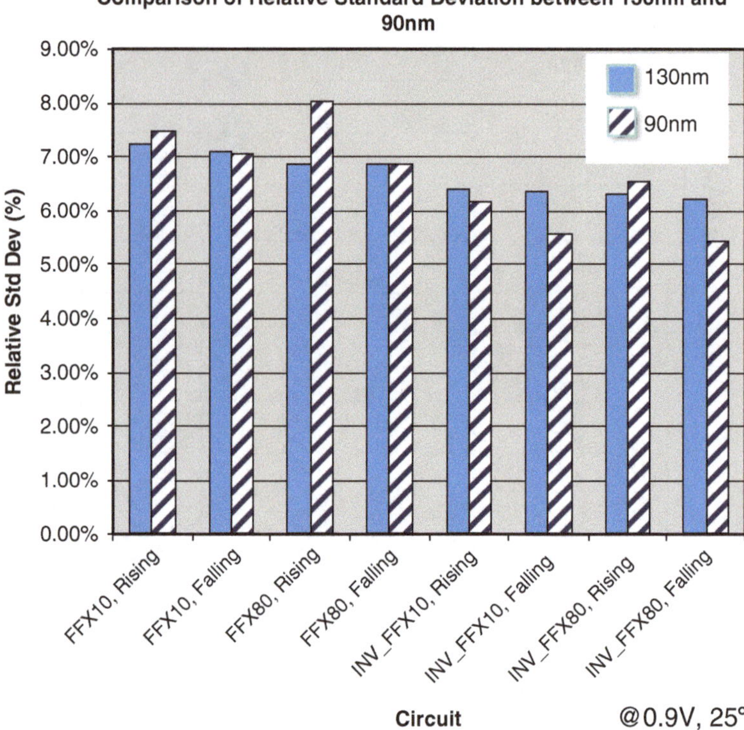

Fig. 6.21 Comparison between 130 wafer 2 and 90 nm wafer at Vdd = 0.9 V – relative standard deviation

Chapter 7
Systematic and Random Variability

Of Measured Results

The total variability observed in the measured data may come from different sources. They may be wafer-to-wafer, die-to-die, intra-die, and may come from systematic or random sources.

Systematic variations come from systematic errors in the fabrication process, and are more dependent, for example, of the position of the die in the wafer. Random variations are more independent of the position, since they come from more unpredictable, normal Gaussian sources (like the position of dopant atoms).

In this chapter, we make use of mathematical methods to try to separate the measured results between systematic and random variability, making possible a better understanding of the results.

7.1 Separation Methods

With the discussed measurement technique, it is possible to measure the overall variability on the wafer. However, for a deeper analysis, it is necessary to make mathematical transformations in the obtained data. Several methods to make the separation between the different components of the variability are present in the literature [7, 18, 45]. In this work, we will focus in how to separate the data between systematic (over the wafer) variability and random residual (within-die, local, or residuals due to imperfection in the measurement) variability.

A simple but widely used method is the moving average. In this method, the measured value in each die is substituted by the average of the value in the die itself with the values of the neighbor dies. If the number of dies is large, the average window can be expanded. We will analyze the results using a 3×3 window (the die with its direct adjacent neighbors) and a 5×5 window (with neighbors up to 2 dies of distance). The drawback of this method is some deterioration at the borders, since we do not have all neighbors available for average calculation.

Another common method is curve fitting. In this method, we take the measured data and apply a linear regression to find the curve that fits it better. The curve can

G. Neuberger et al., *Protecting Chips Against Hold Time Violations Due to Variability*,
DOI 10.1007/978-94-007-2427-3_7,

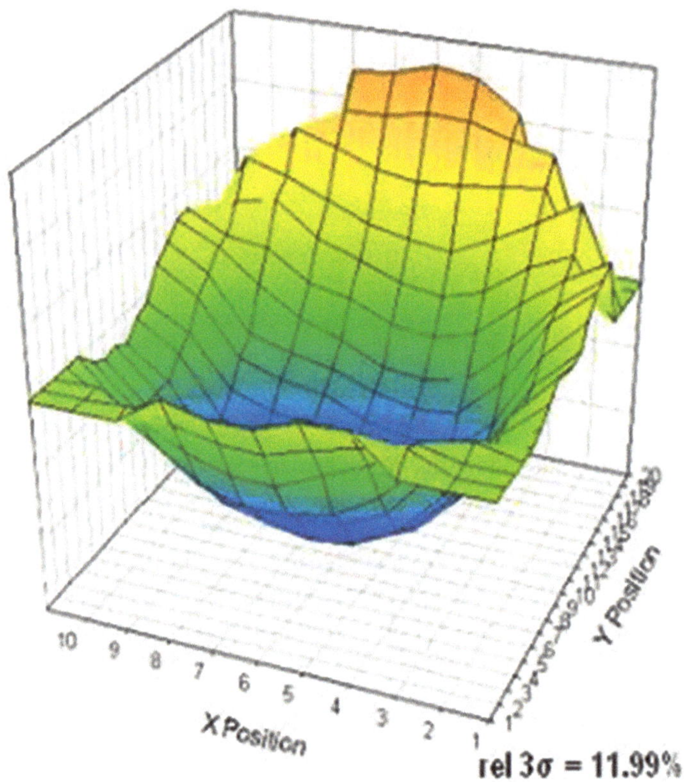

Fig. 7.1 Variability curves in 130 nm wafer: systematic variability using method Moving Average with 3×3 window

be a paraboloid, a plane, a Gaussian, among others, depending on specific issues of the fabrication process. This is a more complex method, and requires a mathematic intensive computation.

7.2 Separation Results

The first step in the analysis was to apply the separation methods described in the previous section in the RO frequency variability. The three methods were compared: moving average with a 3×3 window, moving average with a 5×5 window, and curve fitting. Figures 7.1, 7.2, 7.3, 7.4, 7.5 and 7.6 show the curves obtained for the 130 nm wafer, while Figs. 7.7, 7.8, 7.9, 7.10, 7.11 and 7.12 show the results for the 90 nm wafer. In 130 nm, the curve obtained was a paraboloid, what could be observed already in the original data. However, in the 90 nm wafer, the original data was very random and difficult to see any systematic dependence, but the

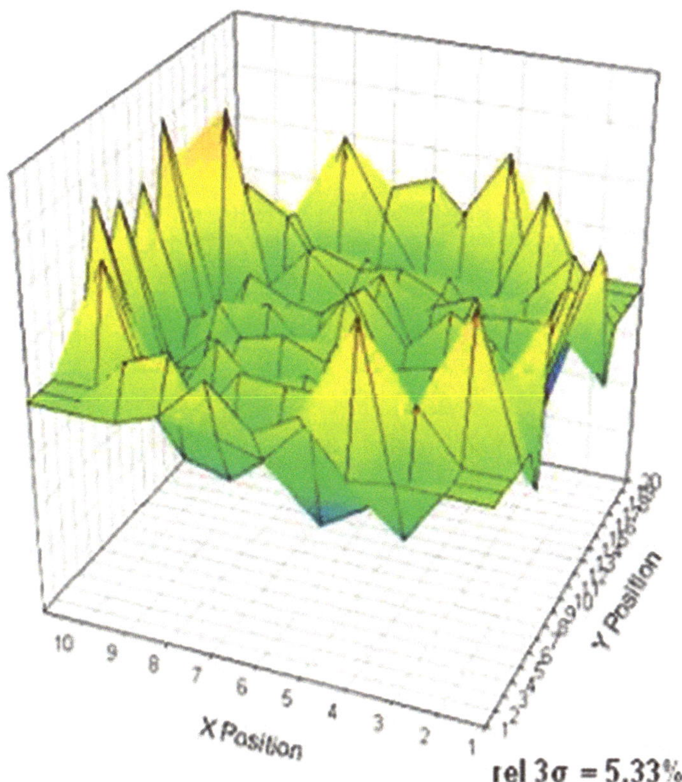

Fig. 7.2 Variability curves in 130 nm wafer: residual variability using method Moving Average with 3×3 window

mathematical methods showed a plane, with ring oscillator frequency increasing from one side of the wafer to the other.

Regarding the numerical results, the standard deviation calculated with the 3×3 moving average method was very close to the one found with the curve fitting method. However, the 5×5 moving average method presented results more than 20 % different from the two other methods, always decreasing systematic variability while increasing the random residuals, showing that a 5×5 window may be too large for the available data, masking part of the systematic variability, and especially leading to a deformation at the corners.

Based on these results, we decided to continue the analysis using only the 3×3 moving average method, due to its simplicity and very close results compared to curve fitting. The final step was to apply the method in the data obtained for the critical clock skew distribution in all circuit configurations. Table 7.1 shows the results of the total measured variability, and the systematic and residual variability calculated with the method, in the 130 nm wafer, while the results for 90 nm are in Table 7.2.

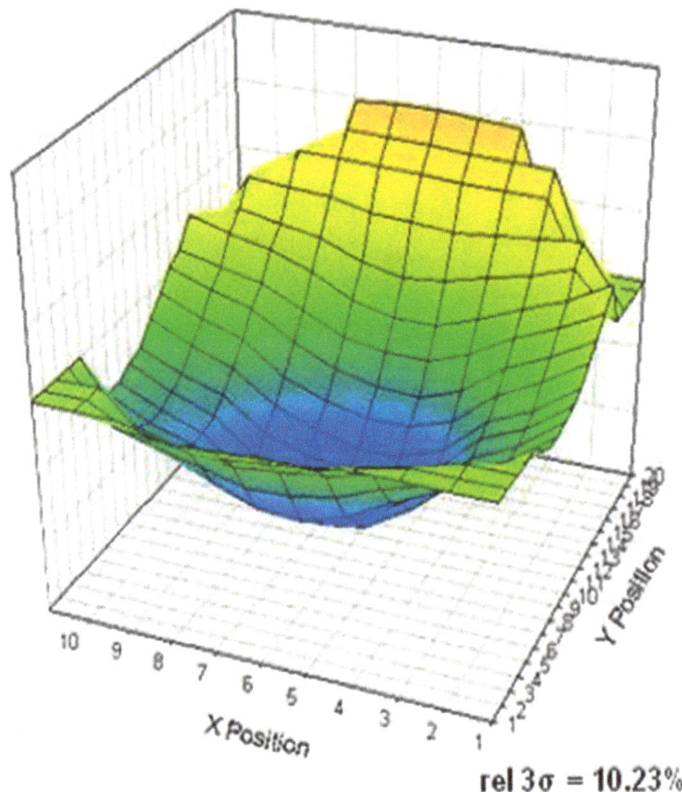

Fig. 7.3 Variability curves in 130 nm wafer: systematic variability using method Moving Average with 5×5 window

Regarding the numerical results, the standard deviation calculated with the 3×3 moving average method was very close to the one found with the curve fitting method. However, the 5×5 moving average method presented results more than 20 % different from the two other methods, always decreasing systematic variability while increasing the random residuals, showing that a 5×5 window may be too large for the available data, masking part of the systematic variability, and especially leading to a deformation at the corners.

Based on these results, we decided to continue the analysis using only the 3×3 moving average method, due to its simplicity and very close results compared to curve fitting. The final step was to apply the method in the data obtained for the critical clock skew distribution in all circuit configurations. Table 7.1 shows the results of the total measured variability, and the systematic and residual variability calculated with the method, in the 130 nm wafer, while the results for 90 nm are in Table 7.2.

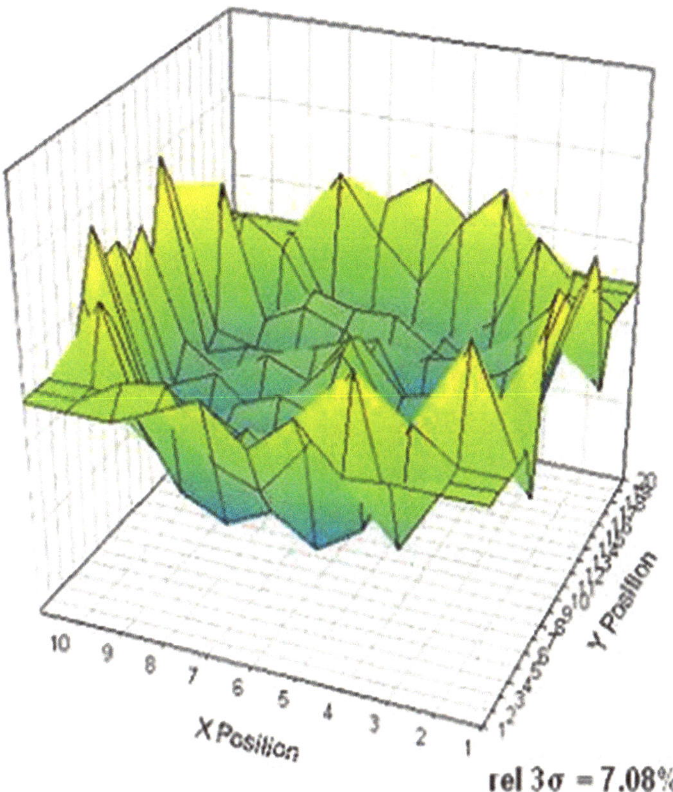

Fig. 7.4 Variability curves in 130 nm wafer: residual variability using method Moving Average with 5×5 window

The results show that the systematic variability is dominant in 130 nm technology. However, with scaling to 90 nm, the residual, probably influenced by the local and within-die variability, is becoming much more important (about two times increase from 130 to 90 nm), while the systematic variability is decreasing due to a better process control.

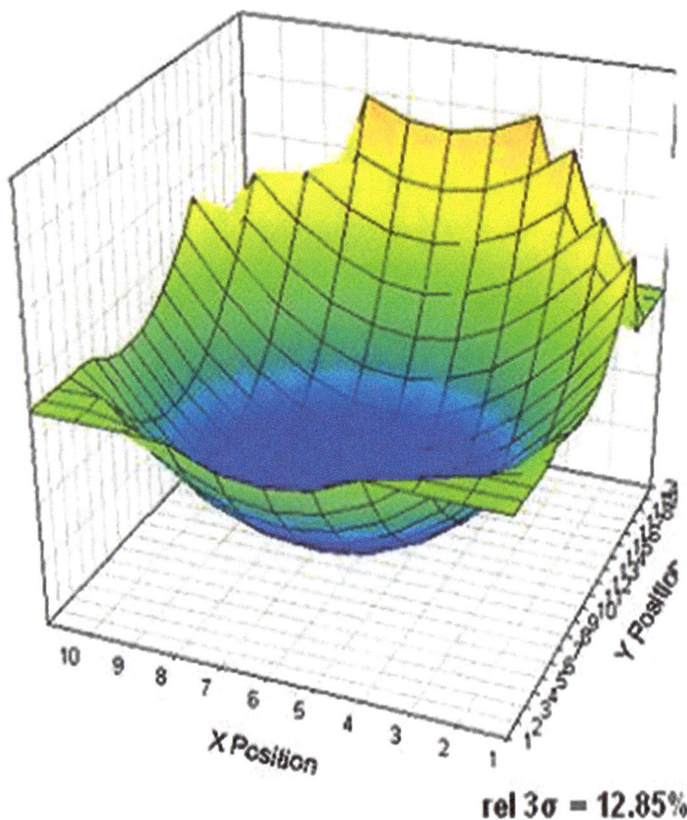

Fig. 7.5 Variability curves in 130 nm wafer: systematic variability using method Curve Fitting

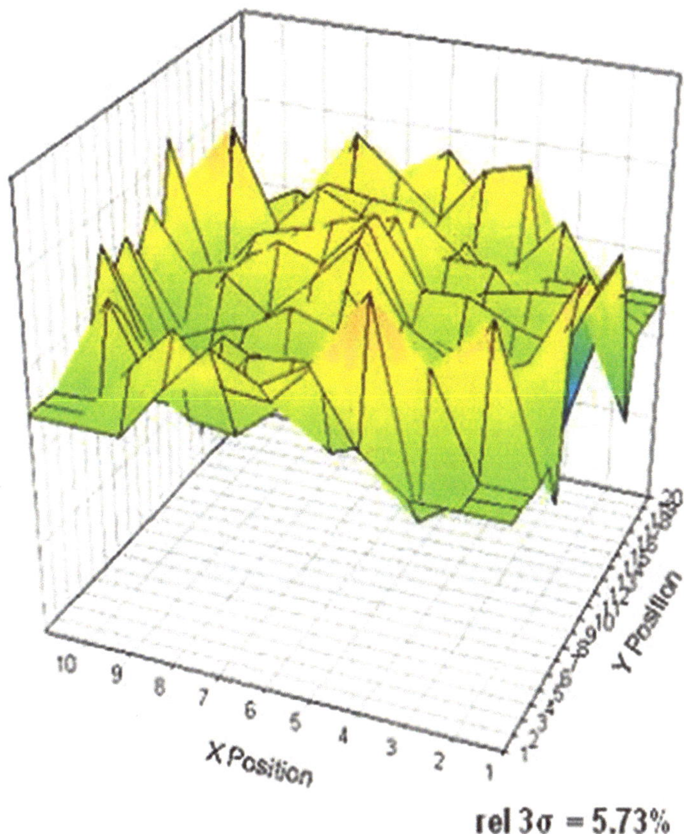

Fig. 7.6 Variability curves in 130 nm wafer: residual variability using method Curve Fitting

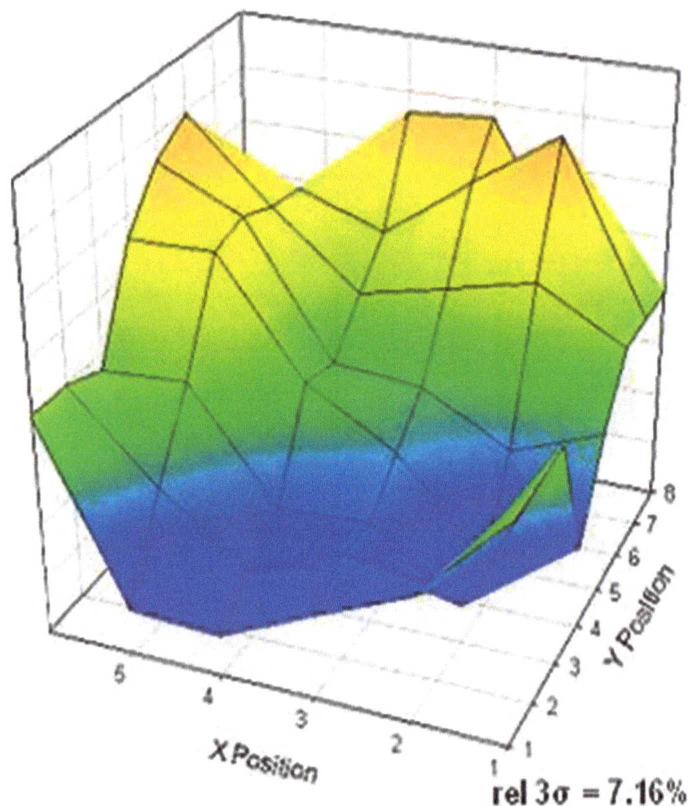

Fig. 7.7 Variability curves in 90 nm wafer: systematic variability using method Moving Average with 3 × 3 window

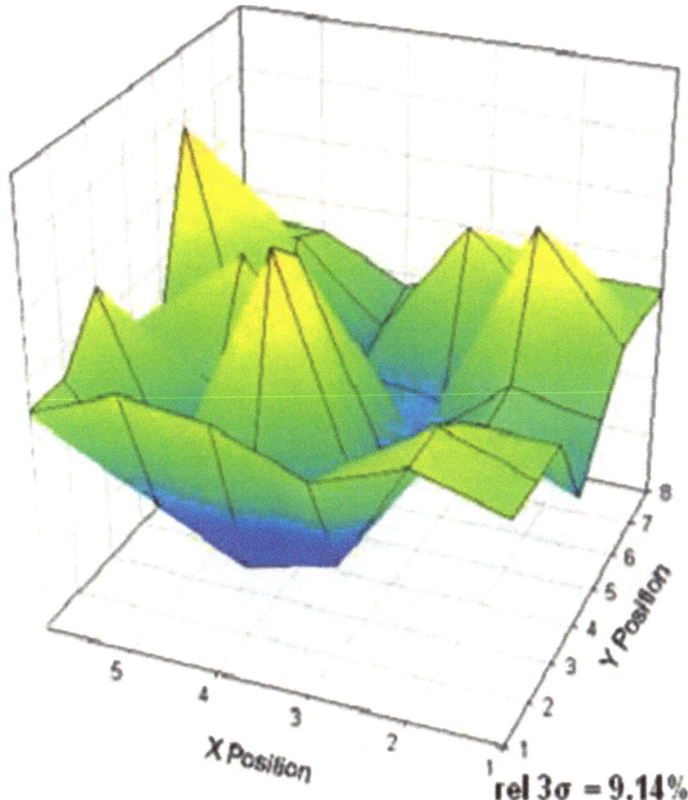

Fig. 7.8 Variability curves in 90 nm wafer: residual variability using method Moving Average with 3×3 window

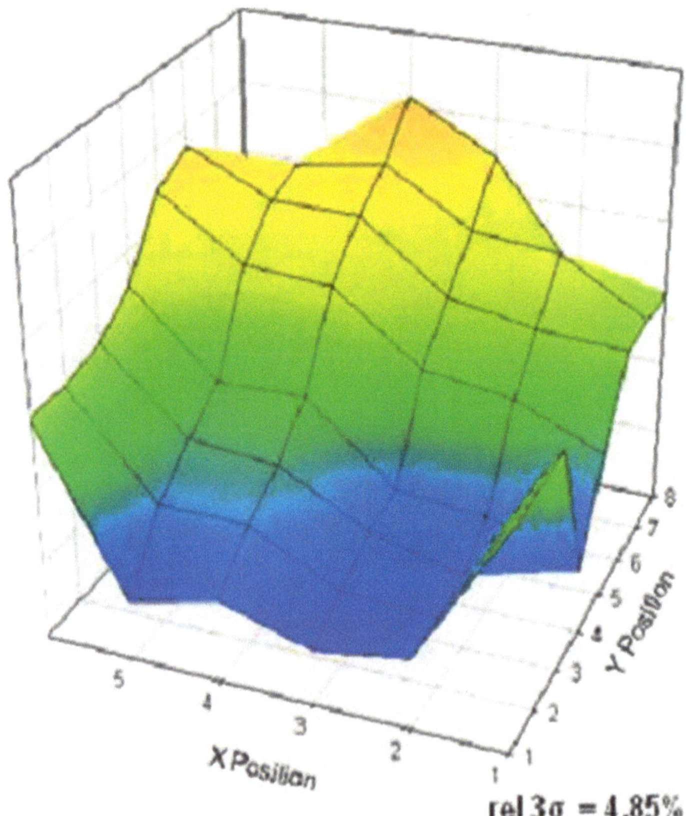

Fig. 7.9 Variability curves in 90 nm wafer: systematic variability using method Moving Average with 5 × 5 window

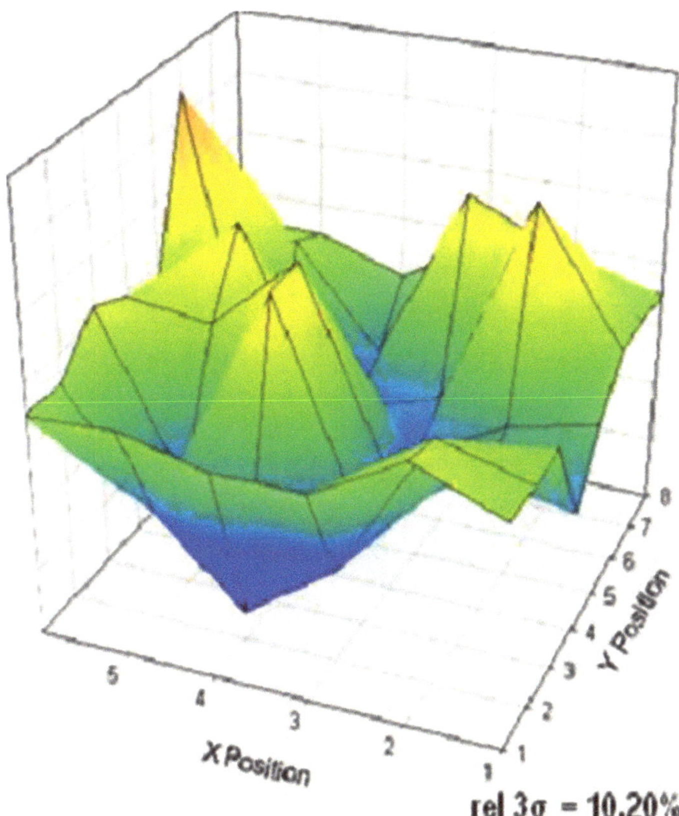

Fig. 7.10 Variability curves in 90 nm wafer: residual variability using method Moving Average with 5 × 5 window

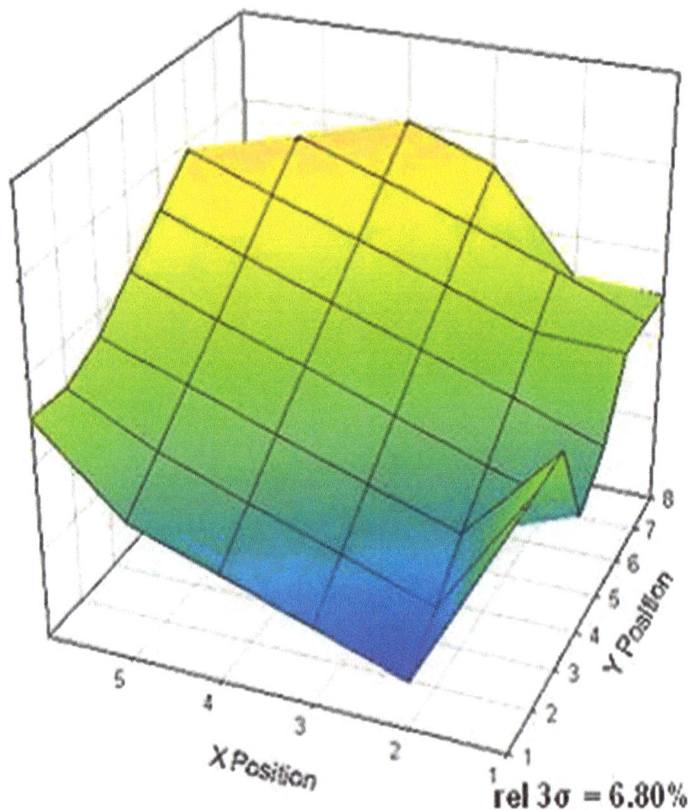

Fig. 7.11 Variability curves in 90 nm wafer: systematic variability using method Curve Fitting

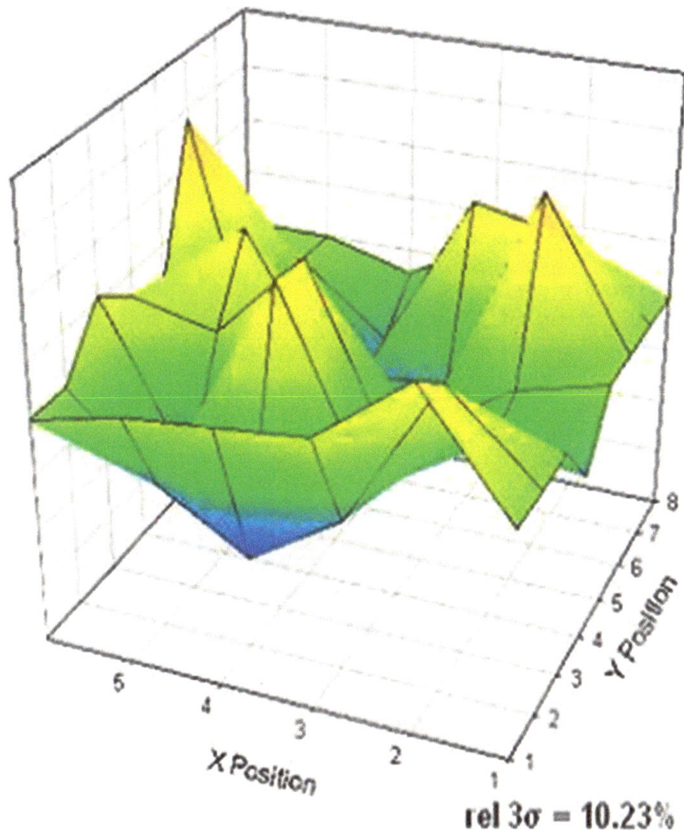

Fig. 7.12 Variability curves in 90 nm wafer: residual variability using method Curve Fitting

Table 7.1 Total, systematic and random residual variability in the critical clock skew using the 3×3 moving average method in 130 nm wafer 2 at V_{DD}=1.5V and T=25°

Circuit	Transition	3σ Total (%)	3σ systematic (%)	3σ residual (%)
Weak FFs,	Rising	14.88	12.12	7.51
No inverters	Falling	13.28	10.91	6.68
Strong FFs,	Rising	13.57	10.42	7.59
No inverters	Falling	12.93	10.06	7.11
Weak FFs,	Rising	13.08	10.68	6.44
6 inverters	Falling	12.04	9.98	5.68
Strong FFs,	Rising	13.13	10.94	5.83
6 Inverters	Falling	12.51	10.35	5.59

Table 7.2 Total, systematic and random residual variability in the critical clock skew using the 3×3 moving average method in 90 nm wafer at V_{DD}=1.32V and T=25°

Circuit	Transition	3σ total (%)	3σ systematic (%)	3σ residual (%)
Weak FFs,	Rising	16.49	8.17	13.19
No inverters	Falling	15.20	6.79	13.49
Strong FFs,	Rising	16.04	8.83	11.49
No inverters	Falling	14.73	8.34	11.03
Weak FFs,	Rising	14.70	7.78	11.67
6 inverters	Falling	14.20	8.49	10.55
Strong FFs,	Rising	14.83	7.74	11.47
6 inverters	Falling	13.67	7.50	10.26

Chapter 8
Normality Tests

Of Measured Results

To evaluate the randomness and check if the measured variability data is a normal Gaussian, mathematical normality tests were performed in the results. There are several tests that are designed to check for normality [14, 28].

In this chapter, the normality tests applied to the measured data of Chap. 6 will be presented, together with the results obtained. These results were presented first in [34].

8.1 The Tests

Different tests for measuring the normality of a set of data are available in the literature. In this book, 3 different tests were chosen to check the normality of the measurements: Wilks-Shapiro test, Anderson-Darling test, and Kurtosis and Skewness analysis. They were chosen because they apply more importance to different aspects, such as the tails or the middle of the distribution.

For the measured data, these tests were applied to data from all test circuits, using the total data, but also to data for systematic and random parts separated. The software used to perform the tests was DataPlot from NIST/Sematech [35].

8.1.1 Wilks-Shapiro Test

The first test used in this work in the data was the Wilks-Shapiro (or Shapiro-Wilk) test [40]. It returns a number called p-value (or W), which may lay between 0 and 1. The larger this number is, more likely is the distribution to be normal. A p-value larger than 0.05 is said to be a normal Gaussian curve at the 95 % confidence level.

In statistics, the Wilks-Shapiro test tests the null hypothesis that a sample $x_1, \ldots, x_n$ came from a normally distributed population. It was published in 1965 by Samuel Shapiro and Martin Wilk.

G. Neuberger et al., *Protecting Chips Against Hold Time Violations Due to Variability*, DOI 10.1007/978-94-007-2427-3_8, © Springer Science+Business Media Dordrecht 2014

The test statistic (Eq. 8.1) is:

$$W = \frac{\left(\sum_{i=1}^{n} a_i x_{(i)}\right)^2}{\sum_{i=1}^{n} \left(x_i - \bar{x}\right)^2} \tag{8.1}$$

where

- $x_{(i)}$ (with parentheses enclosing the subscript index i) is the ith order statistic, i.e., the ith-smallest number in the sample;
- $\bar{x} = \left(x_1 + \ldots + x_n\right) / n$ is the sample mean;
- the constants a_i are given by

$$\left(a_1, \ldots, a_n\right) = \frac{m^T V^{-1}}{\left(m^T V^{-1} V^{-1} m\right)^{1/2}}$$

where

$$m = \left(m_1, \ldots.., m_n\right)^T$$

and $m_1, \ldots, m_n$ are the expected values of the order statistics of independent and identically-distributed random variables sampled from the standard normal distribution, and V is the covariance matrix of those order statistics.

The user may reject the null hypothesis if W is too small.

Monte Carlo simulations studies have indicated that the Wilks-Shapiro test has good power properties for a wide range of alternative distributions.

8.1.2 Anderson-Darling Test

Another common test is the Anderson-Darling normality test [1]. The result of this test is a number larger than 0. But now, the smaller this number, more likely is the distribution to be normal. It is considered that a value smaller than 0.787 relates to a normal Gaussian distribution at the 95 % confidence level.

The Anderson-Darling test, named after Theodore Wilbur Anderson, Jr. and Donald A. Darling, who invented it in 1952, is one of the most powerful statistics for detecting most departures from normality. The Anderson-Darling test assesses whether a sample comes from a specified distribution. The formula for the test statistic A to assess if data $\left\{X_1 < \ldots < X_n\right\}$ (note that the data must be put in order) comes from a distribution with cumulative distribution function (CDF) Φ is (Eq. 8.2):

$$A^2 = -n - \frac{1}{n} \sum_{i=1}^{n} (2i-1)\left(\ln \Phi\left(Y_i\right) + \ln\left(1 - \Phi\left(Y_{n+1-i}\right)\right)\right) \tag{8.2}$$

where

- $Y_i = \dfrac{X_i - \bar{X}}{\sigma}$;

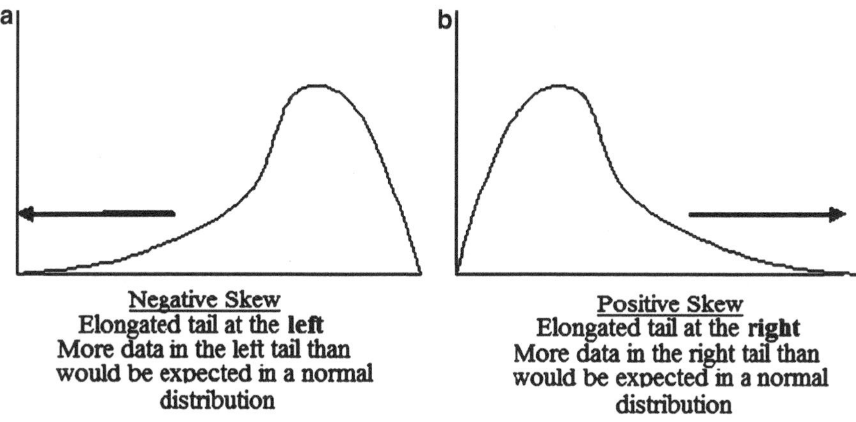

Fig. 8.1 Comparison between 130 nm wafer 2 and 90 nm wafer at Vdd = 1.2V – (**a**) average critical skew and (**b**) relative standard deviation

- $\bar{x} = (x_1 + \ldots + x_n)/n$ is the sample mean;
- σ is the standard deviation.

In comparisons of power, [44] found A-D test to be one of the best empirical distribution function (EDF) statistics for detecting most departures from normality. The only statistic close was the Wilks-Shapiro statistic.

8.1.3 *Kurtosis and Skewness Analysis*

An alternative way to check the normality is to calculate the kurtosis and the skewness of the data [24]. Kurtosis is based on the size of a distribution's tails. A kurtosis of about 3 means a distribution very close to a normal distribution. Skewness is the measure of the asymmetry of the distribution. A normal distribution should have this value equal to 0.

In probability theory and statistics, kurtosis is a measure of the "peakedness" of the probability distribution of a real-valued random variable. Higher kurtosis means that most of the variance is due to infrequent extreme deviations, as opposed to frequent modestly sized deviations.

The kurtosis is normally calculated as the fourth standardized moment, defined as μ_4 / σ^4, where μ_4 is the fourth moment about the mean and σ is the standard deviation. A high kurtosis distribution has a sharper "peak" and fatter "tails", while a low kurtosis distribution has a more rounded peak with wider "shoulders".

In probability theory and statistics, skewness is a measure of the asymmetry of the probability distribution of a real-valued random variable.

Consider the distribution in the Fig. 8.1. The bars on the right side of the distribution taper differently than the bars on the left side. These tapering sides are called tails, and they provide a visual means for determining which of the two kinds of skewness a distribution has:

1. Positive skew: The right tail is the longest; the mass of the distribution is concentrated on the left of the figure. The distribution is said to be right-skewed.
2. Negative skew: The left tail is the longest; the mass of the distribution is concentrated on the right of the figure. The distribution is said to be left-skewed.

Skewness, the third standardized moment, is written as γ_1 and defined as μ_3/σ^3, where μ_3 is the third moment about the mean and σ is the standard deviation.

8.2 Normality Test Results

The next step was to apply the normality tests to the set of data. All 8 test paths and also the ring oscillator frequency were tested, using the methods described previously. Table 8.1 shows the results of these tests applied to the ring oscillator frequency data from the 130 nm wafer, while Tables 8.2, 8.3, 8.4 and 8.5 show the results for the different test paths.

Analyzing the results, it is possible to see some similarities between the results of different test circuits. First, Wilks-Shapiro and Anderson-Darling tests produced consistent conclusions in all cases. In all cases, random variability is more normal

Table 8.1 Wilks-Shapiro and Anderson-Darling normality tests, kurtosis and skewness for total, systematic and random residual variability for frequency data in 130 nm wafer 2 at $V_{DD} = 1.5$ V and $T = 25°$

	Frequency		
	Total	Systematic	Residual
W-S p-value	0.00013	0.00042	0.000056
Conclusion	Reject	Reject	Reject
A-D value	1.806	1.084	2.376
Conclusion	Reject	Reject	Reject
Kurtosis	2.393	2.565	3.105
Skewness	0.456	0.417	0.685

Table 8.2 Wilks-Shapiro and Anderson-Darling normality tests, kurtosis and skewness for total, systematic and random residual variability for test path of weak FFs, no inverters, in 130 nm wafer 2 at $V_{DD} = 1.5$V and $T = 25°$

	Weak FFs, no inv's, rising			Weak FFs, no inv's, falling		
	Total	Syst.	Residual	Total	Syst.	Residual
W-S p-value	0.1907	0.00109	0.447	0.114	0.00101	0.269
Conclusion	Accept	Reject	Accept	Accept	Reject	Accept
A-D value	0.464	0.987	0.25	0.673	1.128	0.672
Conclusion	Accept	Reject	Accept	Accept	Reject	Accept
Kurtosis	2.396	2.317	2.89	2.434	2.313	3.062
Skewness	−0.134	−0.25	0.00372	−0.188	−0.356	0.167

Table 8.3 Wilks-Shapiro and Anderson-Darling normality tests, kurtosis and skewness for total, systematic and random residual variability for test path of strong FFs, no inverters, in 130 nm wafer 2 at V_{DD}=1.5V and T=25°

	Strong FFs, no inv's, rising			Strong FFs, no inv's, falling		
	Total	Syst.	Residual	Total	Syst.	Residual
W-S p-value	0.0259	0.0023	0.0783	0.00115	0.00028	0.957
Conclusion	Reject	Reject	Accept	Reject	Reject	Accept
A-D value	0.941	0.919	0.682	1.284	1.299	0.208
Conclusion	Reject	Reject	Accept	Reject	Reject	Accept
Kurtosis	2.224	2.526	2.197	2.111	2.346	2.786
Skewness	–0.149	–0.359	–0.00355	–0.227	–0.344	–0.0489

Table 8.4 Wilks-Shapiro and Anderson-Darling normality tests, kurtosis and skewness for total, systematic and random residual variability for test path of weak FFs, 6 inverters, in 130 nm wafer 2 at V_{DD}=1.5 V and T=25°

	Weak FFs, 6 inv's, rising			Weak FFs, 6 inv's, falling		
	Total	Syst.	Residual	Total	Syst.	Residual
W-S p-value	0.0247	0.0055	0.109	0.00201	0.00158	0.109
Conclusion	Reject	Reject	Accept	Reject	Reject	Accept
A-D value	0.674	0.696	0.596	1.015	1.006	0.731
Conclusion	Accept	Accept	Accept	Reject	Reject	Accept
Kurtosis	2.595	2.571	3.124	2.419	2.459	3.72
Skewness	–0.335	–0.316	–0.392	–0.358	–0.396	–0.0778

Table 8.5 Wilks-Shapiro and Anderson-Darling normality tests, kurtosis and skewness for total, systematic and random residual variability for test path of strong FFs, 6 inverters, in 130 nm wafer 2 at V_{DD}=1.5 V and T=25°

	Strong FFs, 6 inv's, rising			Strong FFs, 6 inv's, falling		
	Total	Syst.	Residual	Total	Syst.	Residual
W-S p-value	0.0169	0.00271	0.0681	0.00074	0.0003	0.111
Conclusion	Reject	Reject	Accept	Reject	Reject	Accept
A-D value	1.009	0.873	0.759	1.451	1.344	0.718
Conclusion	Reject	Reject	Accept	Reject	Reject	Accept
Kurtosis	2.351	2.392	2.533	2.245	2.281	2.877
Skewness	–0.259	–0.306	–0.27	–0.329	–0.336	–0.329

than the equivalent systematic variability, while the total data is located somewhere between them. The only exception is the ring oscillator frequency, but the reason is that it uses a large number of inverters (17), made of larger transistors than the other cases, thus minimizing the random variability.

In all cases except for the ring oscillator frequency, the random variability is considered normal in the conclusion of the tests, while systematic variability is not normal (as expected, since it has a strong spatial dependence from the middle to the

Table 8.6 Wilks-Shapiro and Anderson-Darling normality tests, kurtosis and skewness for total, systematic and random residual variability for frequency data in 90 nm wafer at $V_{DD}=1.32$ V and T=25°

	Frequency		
	Total	Systematic	Residual
W-S p-value	0.2876	0.03704	0.9734
Conclusion	Accept	Reject	Accept
A-D value	0.598	0.753	0.148
Conclusion	Accept	Reject	Accept
Kurtosis	2.367	1.664	2.453
Skewness	0.351	0.186	0.132

Table 8.7 Wilks-Shapiro and Anderson-Darling normality tests, kurtosis and skewness for total, systematic and random residual variability for test path of weak FFs, no inverters, in 90 nm wafer at $V_{DD}=1.32$V and T=25°

	Weak FFs, no inv's, rising			Weak FFs, no inv's, falling		
	Total	Syst.	Residual	Total	Syst.	Residual
W-S p-value	0.8273	0.2245	0.2324	0.3772	0.2909	0.3593
Conclusion	Accept	Accept	Accept	Accept	Accept	Accept
A-D value	0.26	0.442	0.436	0.451	0.466	0.397
Conclusion	Accept	Accept	Accept	Accept	Accept	Accept
Kurtosis	2.978	2.266	2.423	2.149	1.908	1.947
Skewness	0.226	–0.433	0.205	0.0393	–0.104	–0.0676

Table 8.8 Wilks-Shapiro and Anderson-Darling normality tests, kurtosis and skewness for total, systematic and random residual variability for test path of strong FFs, no inverters, in 90 nm wafer at $V_{DD}=1.32$V and T=25°

	Weak FFs, no inv's, rising			Weak FFs, no inv's, falling		
	Total	Syst.	Residual	Total	Syst.	Residual
W-S p-value	0.8728	0.4015	0.5593	0.6162	0.1909	0.3611
Conclusion	Accept	Accept	Accept	Accept	Accept	Accept
A-D value	0.242	0.476	0.306	0.274	0.469	0.412
Conclusion	Accept	Accept	Accept	Accept	Accept	Accept
Kurtosis	2.224	2.024	2.498	2.568	2.276	2.288
Skewness	0.143	0.14	0.229	0.386	0.435	0.169

corners of the wafer). The total data is considered normal in the cases with weak FFs and no inverters, since these are the cases where random variability prevails, while it is not normal in other cases. In the case where we have large FFs and inverters, the role of random (local) variability is diminished and the role of systematic variability increases.

Analyzing the kurtosis and skewness results, it is possible to see that in all cases, they are close to the values of 3 and 0, respectively, as expected for normal Gaussian curves.

Now, Tables 8.6, 8.7, 8.8, 8.9 and 8.10 show the tests results for 90 nm.

These results repeat the tendency found in the 130 nm results, of random variability presenting higher values of normality compared to systematic variability. But now,

Table 8.9 Wilks-Shapiro and Anderson-Darling normality tests, kurtosis and skewness for total, systematic and random residual variability for test path of weak FFs, 6 inverters, in 90 nm wafer at $V_{DD} = 1.32$ V and T = 25°

	Weak FFs, no inv's, rising			Weak FFs, no inv's, falling		
	Total	Syst.	Residual	Total	Syst.	Residual
W-S p-value	0.7571	0.0785	0.8608	0.6402	0.0048	0.8844
Conclusion	Accept	Accept	Accept	Accept	Reject	Accept
A-D value	0.268	0.789	0.292	0.209	1.369	0.198
Conclusion	Accept	Reject	Accept	Accept	Reject	Accept
Kurtosis	3.321	1.898	3.048	2.704	2.008	2.33
Skewness	0.0125	–0.328	0.375	–0.416	–0.591	0.054

Table 8.10 Wilks-Shapiro and Anderson-Darling normality tests, kurtosis and skewness for total, systematic and random residual variability for test path of strong FFs, 6 inverters, in 90 nm wafer at $V_{DD} = 1.32$ V and T = 25°

	Weak FFs, no inv's, rising			Weak FFs, no inv's, falling		
	Total	Syst.	Residual	Total	Syst.	Residual
W-S p-value	0.3415	0.0318	0.3319	0.7577	0.03263	0.8858
Conclusion	Accept	Reject	Accept	Accept	Reject	Accept
A-D value	0.309	0.823	0.337	0.396	0.884	0.191
Conclusion	Accept	Reject	Accept	Accept	Reject	Accept
Kurtosis	2.152	2.075	2.224	2.529	1.703	2.529
Skewness	–0.303	–0.496	–0.178	–0.153	–0.262	0.0189

the normality values are much larger in all cases, meaning that both random and systematic variability are becoming more normal with the technology advancement.

In all cases, random variability and the total measured data are accepted to be normal Gaussian distributions as conclusion of the tests. However, systematic variability is now considered normal in all cases with no inverters, while it is not normal in most cases with 6 inverters. Since the inverters make an average of the random residual variability between themselves, hiding the normality, it can be considered that in 90nm, systematic variability is normal.

The results of kurtosis and skewness are again close to the values of 3 and 0 of normal distributions.

Chapter 9
Probability of Hold Time Violations

Of Short Logic Paths

In previous chapters, the methodology to measure the value of the race immunity of a FF was presented. Although this data is important by itself, it is more meaningful when put together with the estimate of the clock skew, and the probability of hold time violations is calculated.

This chapter will show the methodology of how to calculate the probability of hold time violations using the race immunity and clock skew, and how to apply it to different scenarios, analyzing the behavior expected when changing different parameters.

9.1 Simulation of Dependence of Race Immunity on Vdd

Before proceeding with the calculation of the probability of hold time violation, it is important to check the dependence of race immunity and clock skew on Vdd. The measurements showed the data for race immunity at 3 different values of Vdd, and now by simulations, we will study this dependence deeply.

The dependence of clock skew on Vdd was shown by [38, 53]. They found that the increase in clock skew is linearly proportional to the decrease in Vdd (i. e., if the Vdd decreases 10 %, the clock skew increases 10 %). This is about the same relationship found between the delay of an inverter and Vdd.

To find the dependence of the race immunity on Vdd, simulations at different Vdd's were performed, together with the simulation of the delay of 7 inverters (because 7 inverters have a delay value very close to the race immunity value in the technologies used in this work). These simulations were made for a commercially available 0.35 μm AMS technology, because unfortunately, the 130 and 90 nm were not available at the time of these simulations. Figure 9.1 shows the results, for Vdd value between 2.2 and 4.0 V, with steps of 0.1 V.

The results show that the delay and the race immunity have similar values at higher voltages, but the difference increases when the voltage is decreased. This means that the race immunity has a stronger dependence on Vdd than the inverter delay.

G. Neuberger et al., *Protecting Chips Against Hold Time Violations Due to Variability*, DOI 10.1007/978-94-007-2427-3_9,

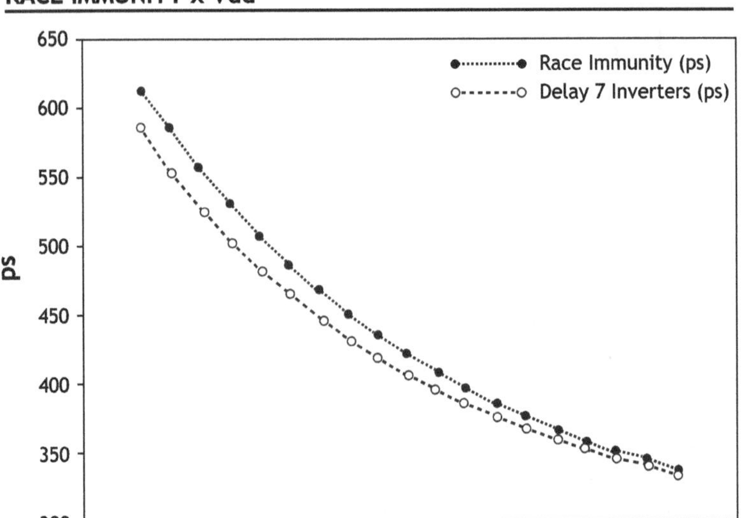

Fig. 9.1 Simulation of race immunity and delay of inverters at different Vdd's

This is justified because the race immunity is composed by both the FF delay, and the FF hold time.

The two curves are closely proportional to 1/(Vdd – VT)x. If we make a fitting of this equation to the curves, we find x to be 0.9 for the race immunity, while being 0.87 for the delay.

Next, it is necessary to estimate the dependence of race immunity standard deviation on Vdd. This was made by Monte Carlo simulations of race immunity. Figure 9.2 shows the results for absolute values of race immunity standard deviation. In this case, only 8 values of Vdd were simulated, because these Monte Carlo simulations take a long time (more than 24 h), and the trend can be well estimated with fewer simulations.

This confirms that the race immunity variability also increases with the decrease of Vdd. However, much more important is the relationship between the race immunity variability and the absolute value. Figure 9.3 shows this relationship (race immunity standard deviation divided by the race immunity value).

Again, the curve is proportional to $1/(Vdd - VT)^x$. Now, the fitting of the curve found x as being exactly equal to 1 for the race immunity standard deviation.

This shows that, with decreasing Vdd, the increase in the variability of race immunity is more significant than the increase in the absolute race immunity value. If the clock skew were always proportional to the race immunity, this result would mean that the probability of hold time violation would increase with the

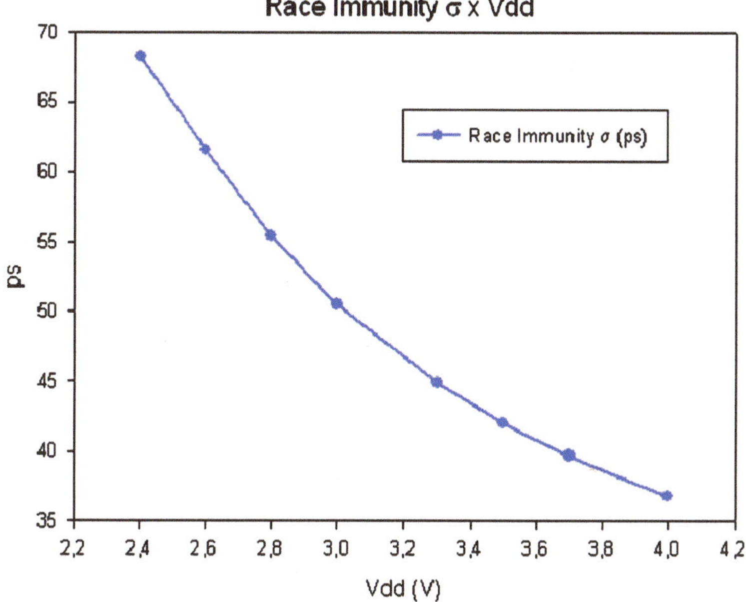

Fig. 9.2 Simulation of absolute race immunity standard deviation at different Vdd's

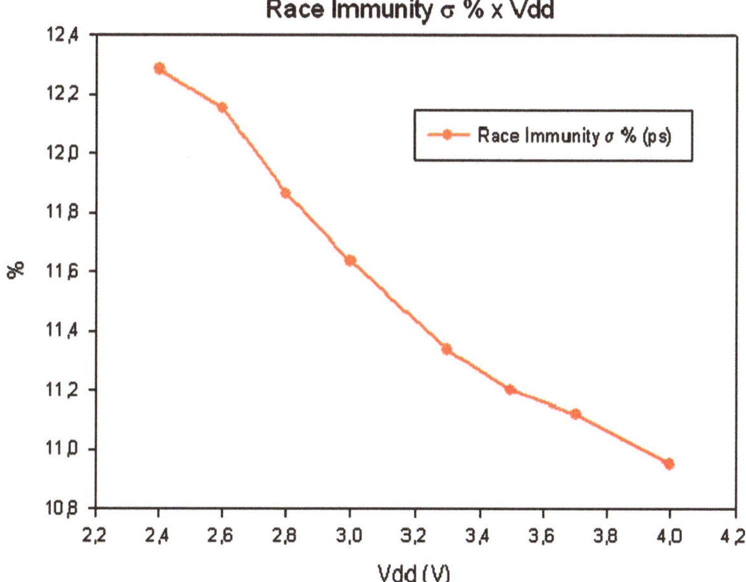

Fig. 9.3 Simulation of relative race immunity standard deviation at different Vdd's

decrease of Vdd. However, the clock skew varies less with Vdd, so the relationship between these results and the hold time violation may be more complicated.

In the next section, we show how to put all this data together, and to calculate the probability of hold time violation.

9.2 Hold Time Violation Probability Using Simulated Data

After the estimation of race immunity average, race immunity standard deviation, and clock skew, it is necessary to calculate the hold time violation probability based on these values. Let us consider that the race immunity is a Gaussian distribution. From Chap. 8 we know that this is not true in all cases, but in most cases it is, and the others can be closely modeled to it. And since in 90 nm the variability becomes more random, and it is expected to continue this trend of randomization and normalization in future technologies, this is a reasonable assumption.

The probability of hold time violation is the probability that the clock skew is higher than the race immunity. For the scenario usually found in the literature, where the clock skew is assumed to be a fixed value, this problem is the same as the problem of the cumulative distribution function, as shown in Fig. 9.4.

The cumulative distribution function (cdf) of a probability distribution, evaluated at a number (lower-case) x, is the probability of the event that a random variable (capital) X with that distribution is less than or equal to x. The cumulative distribution

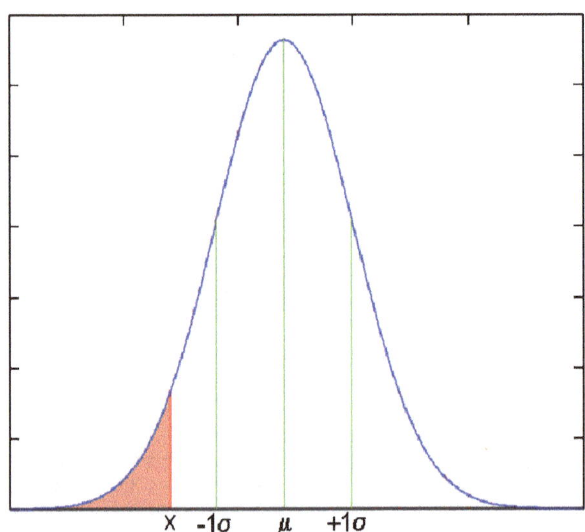

Fig. 9.4 Calculation of hold time violation probability (cumulative distribution function)

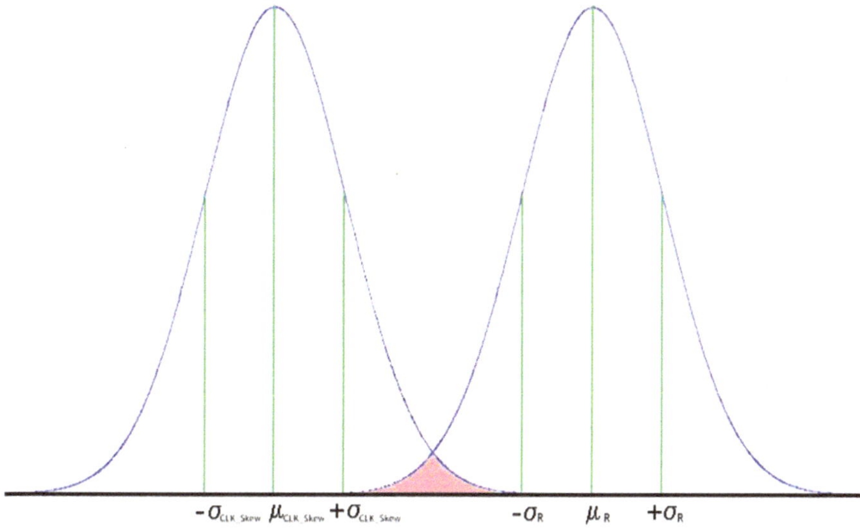

Fig. 9.5 Calculation of hold time violation probability considering clock skew as a normal distribution

function of the normal distribution is expressed in terms of the density function as follows (Eq. 9.1):

$$F(x;\mu,\sigma) = \frac{1}{\sigma\sqrt{2\pi}} \int_{-\mu}^{x} \exp\left(-\frac{(u-\mu)^2}{2\sigma^2} \right) du \tag{9.1}$$

The cdf can also be expressed in terms of a special function called the error function, as (Eq. 9.2):

$$F(x;\mu,\sigma) = \frac{1}{2}\left[1 + erf\left(\frac{x-\mu}{\sigma\sqrt{2}} \right) \right] \tag{9.2}$$

To adapt the problem to the hold time violation probability, we only need to consider x as the expected clock skew, μ as the race immunity average, and σ as the race immunity standard deviation. This method is valid if the clock skew is considered to be a constant value, usually the worst-case clock skew. This corresponds to the methodology usually found in the literature.

However, with the increasing impact of variability on circuit parameters, it may become more appropriate to consider the clock skew to be a random variable, and not a constant value. In this approach, the clock skew may be modeled as a normal distribution. In this case, we must evaluate the probabilities for the race immunity value (normally distributed) being smaller than the clock skew (also a normally distributed random variable). This is the convolution of the two Gaussians, as approximately represented in Fig. 9.5.

To calculate the probability numerically, we will consider that the two distributions are independent normal random variables. This is a valid assumption, since they come mostly from different sources: the race immunity depends most on the variability of transistors and cells, while the clock skew depends most on the variability of global interconnections. Thus, they are weakly correlated. Using this assumption, Eq. 9.3 can be used to calculate the probability of hold time violation [42]:

$$F\left(\mu_R, \sigma_R, \mu_{CLK\_Skew}, \sigma_{CLK\_Skew}\right) = \frac{1}{2}\left[1 + erf\left(\frac{\mu_{CLK\_Skew} - \mu_R}{\sqrt{\left(\sigma_R^2 + \sigma_{CLK\_Skew}^2\right)}\sqrt{2}}\right)\right] \tag{9.3}$$

Although there is no data available on the distribution of clock skew, we may look into some hypothetical cases. Worst-case values are available in the literature, and the most usual assumption in the literature is to assume the worst case value as being mean plus three standard deviations. The hypothetical case studies may then be: (i) to consider the clock skew to be a constant value equal to the worst case presented in literature; (ii) to consider the clock skew to be a random variable, with mean equal to 90 % of worst case and 3*sigma equal to 10 % the of worst case; (iii) to considered the clock skew to be a random variable, with mean equal to 80 % and 3*sigma equal to 20 % the of worst case value.

Now, the presented method is applied to the simulated data of the previous section of this chapter, at different Vdd's. We considered the clock skew in two different cases: the first is that clock skew would be equal to 20 % of the delay of the 7 inverters; and the second case would be equal to 100 % of this same delay.

In the first case, the hold time violation probability was too small at any Vdd (less than 1×10^{-12} in all cases) to be significant to cause any hold time violation. The second case showed more interesting results (Fig. 9.6).

The results show that the probability of violation decreases when Vdd is decreased. This means that the influence of the difference between clock skew and race immunity values is larger than the influence of the increase in race immunity variability.

In this case, the variation of the probability considering different clock skew models was considered. The models were the mentioned previously: the clock skew average and three times standard deviation being: 100 % and 0 %, 90 % and 10 %, or 80 % and 20 % of the worst case clock skew value. Figure 9.7 presents the results for these different clock skew models.

The results show that the dependence on Vdd remains the same (probability increases when Vdd is higher), although in different values. When the clock skew model is 90 % + 10 %, the probability value is about half of the 100 % case. When the model is 80 % + 20 %, the probability is also again about half of the previous one (90 % + 10 %).

The next section presents the calculation of violation probability for different cases, using the real measured data.

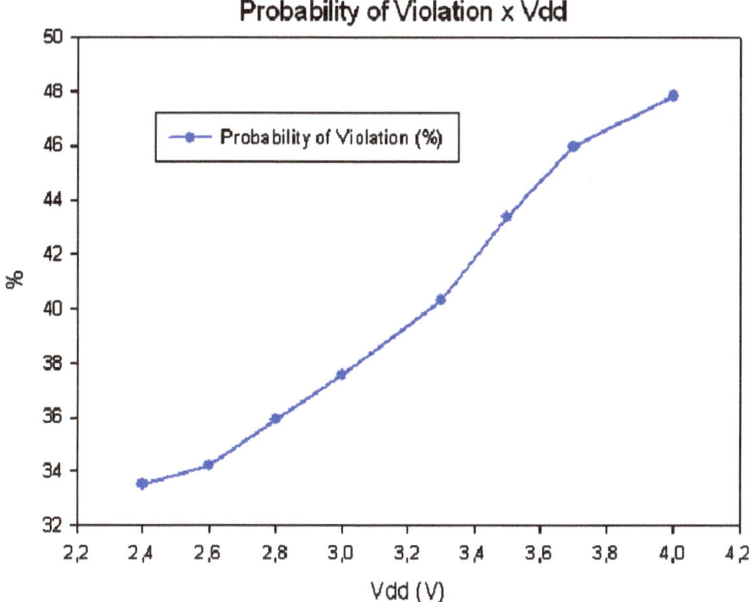

Fig. 9.6 Probability of hold time violation using simulated data

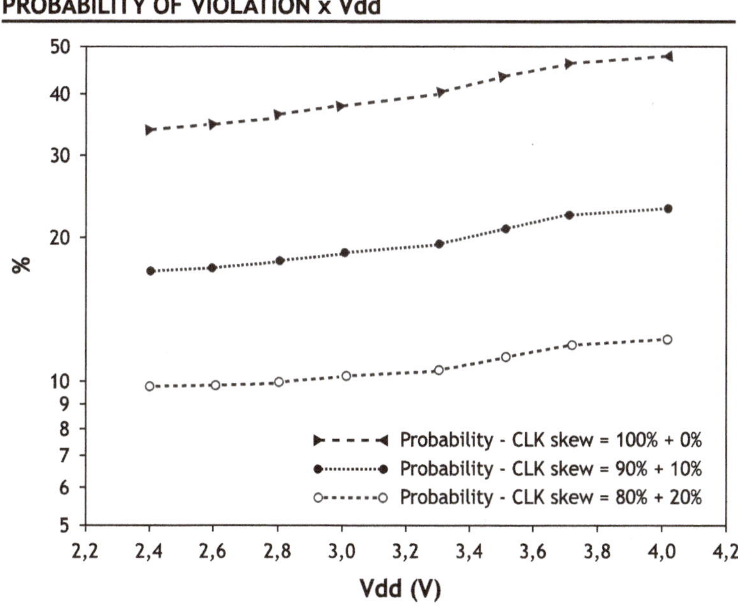

Fig. 9.7 Probability of hold time violation using simulated data, for different clock skew models

9.3 Hold Time Violation Probability Using Measured Data

After the analysis of the potential problem of hold time violations using the simulated data in previous sections, it is necessary to use the real measured data to confirm the results at different scenarios.

For this analysis, we use our measured data of race immunity in both 130 and 90 nm, at all Vdd's used in the measurements, and in the two FF strengths used.

The clock skew estimation considered is as in [32]. It estimates the trend of clock skew between 180 and 50 nm technologies. For example, it says that the worst-case clock skew in 180 nm technology is 16.21 % of the clock period, while at 50 nm technology the clock skew will be 27.96 % of the clock period. In the probability calculations performed in this book, the case where the clock skew average is equal to 100 % of the worst-case total clock skew and sigma is 0 % of the total will be considered.

This section will analyze the trend in the violation probability based on technology, FF strength, Vdd, and padding (placement of buffers or inverters to decrease the violation probability). Finally, also the dependence when using different clock skew models is analyzed.

All these data are for the rising edge violations, since they are worse than the falling violations (rising race immunity is always smaller than falling race immunity), and thus the bottleneck of the problem.

9.3.1 Dependency on Technology

The first trend studied was the technology scaling. The probability of hold time violation was calculated for both 130 and 90 nm, at nominal supply voltage (1.5 V for 130 nm, 1.32 V for 90 nm). The probability was also calculated for both weak (FF_1x) and strong (FF_8x) FFs. Figure 9.8 summarizes the results.

The results show that for both types of FFs, there is a dramatic increase in the probability of violation. For the weak FF, the probability increases from 0.0000822 % in 130 nm to 27.96 % in 90 nm (more than 100,000 times increase), while for the strong FF, it increases from 0.0765 % to 72.26 % (almost 1,000 times). These very high values (more than 10 % in 90 nm) are due principally to the fact that the race immunity decreases significantly (about 30 %) between technologies, while the clock skew does not decrease in the same scale (since it is percentually increasing compared to the clock period).

The results show a trend that is unsustainable, since they show that none circuit would work in 90 nm in this scenario. This analysis is probably overestimated, since it uses a clock skew estimation as a fixed worst-case value. It also does not consider technology developments that already can be used at 90 nm, as FFs with increased race immunity, padding, and a clock skew better controlled (having a smaller value than predicted).

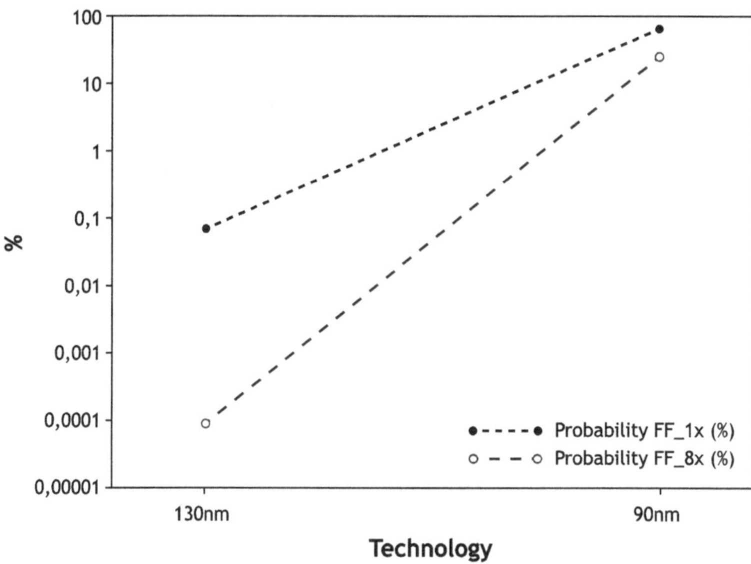

Fig. 9.8 Probability of violation for different technologies

These results would mean that none circuits would be working at recent technologies. However, the use of padding in short paths is already common to achieve correct operation, and these probabilities are for a hypothetical short path with no logic in between.

Nonetheless, these results confirm that the problem is highly increasing with technology scaling and must be treated carefully.

9.3.2 Dependency on Flip-Flop Strength

The next analysis was the dependence of the probability on FF strength. We measured the race immunity for a weak, slower FF (FF_1x) and a strong, faster FF (FF_8x). Considering the same clock skew in both cases, the Fig. 9.9 shows the results for 130 nm, in all the three different voltages of measurements.

The trend in all 3 voltages was the same: the probability of violation increases in stronger, faster FFs. This is expected, since the faster FFs have smaller race immunities. More important is how much is this increase. From Chap. 6, we know that the race immunity of FF_8x is just 12 % smaller than the race immunity of FF_1x. However, the increase in the probability is almost 1,000 times (from 0.0000822 % to 0.0765 %) at 1.5 V, 500 times (from 0.00011 % to 0.0555 %) at 1.2 V, and 75 times (from 0.00825 % to 0.6040 %) at 0.9 V.

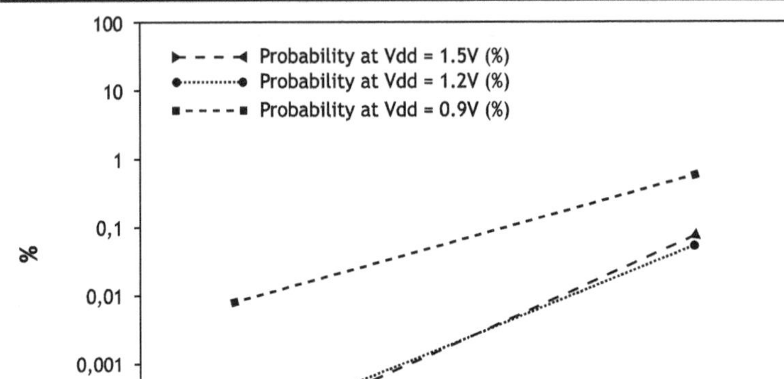

Fig. 9.9 Probability of violation for different FFs in 130 nm

Now, Fig. 9.10 shows the results for 90 nm.

The results show the same trend of increase of probability when the FF is faster. However, since the probabilities are already high, the increase is not so steep.

It is important to determine if the clock skew scales in the same way as the internal race immunity to check if the hold time violation probability changes. However, it is reported by the industry that decreasing Vdd can make some circuits showing violations to work again. It is 2.5 times (from 27.96 % to 72.26 %) at 1.32 V, 2.8 times (from 22.27 % to 63.89 %) at 1.2 V, and 3.2 times (from 11.67 % to 38.07 %) at 0.9 V.

These results lead to the conclusion that the use of faster and stronger FFs should be avoided in fast logic paths. However, this is not always possible (the output of a FF may go to a critical path and a fast path at same time, and then the use of a faster FF would be preferable to decrease the clock period), and then another solution must be used.

9.3.3 Dependency on Vdd

The next analysis is the dependence of the probability on Vdd. The clock skew was considered to remain percentually constant to the clock period when the Vdd is changed. Figure 9.11 presents the results for both FFs in 90 nm technology.

The probability of violation increases when Vdd is increased, for both FFs. This increase is from 11.67 % at 0.9 V to 22.27 % at 1.2 V and 27.96 % at 1.32 V

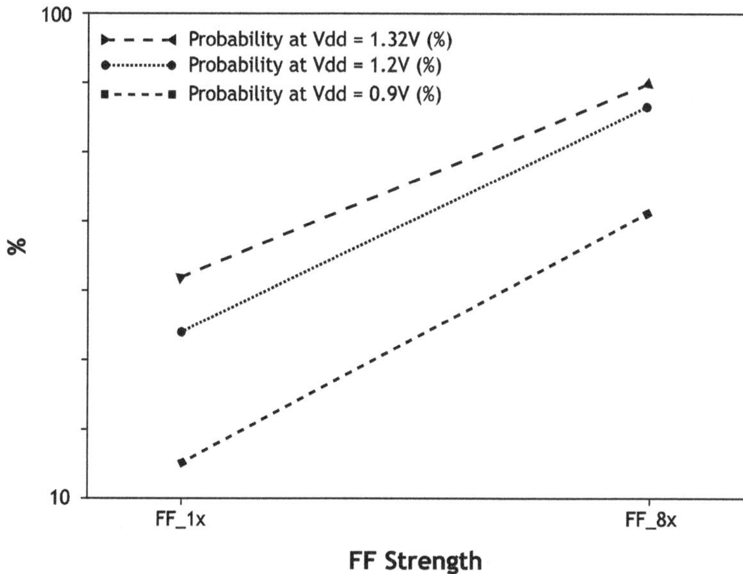

Fig. 9.10 Probability of violation for different FFs in 90 nm

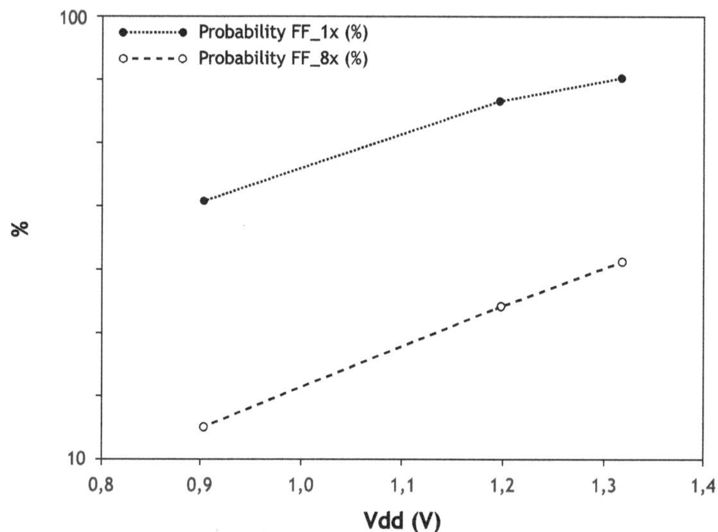

Fig. 9.11 Probability of violation for different Vdd's in 90 nm

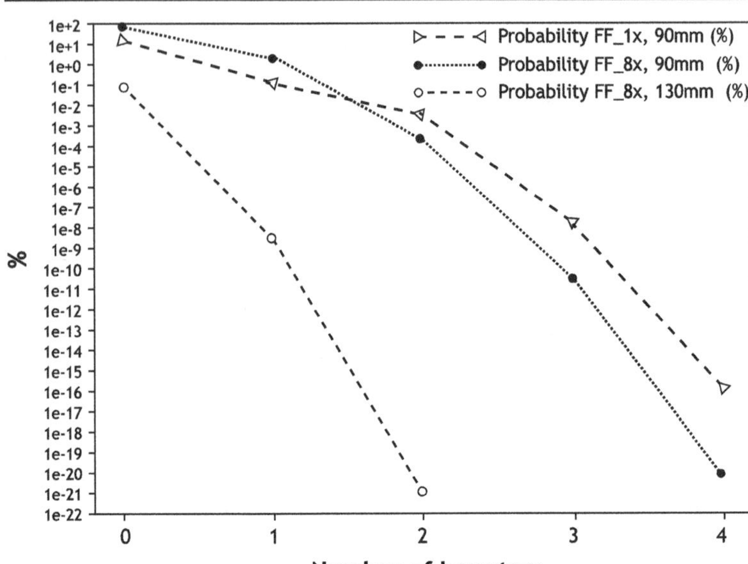

Fig. 9.12 Probability of violation for different padding

for FF_1x, and from 38.07 % at 0.9 V to 63.89 at 1.2 V and 72.26 % at 1.32 V for FF_8x.

In 130 nm technology, the probabilities of violation were very small, and it was not possible to observe the trend (the probability sometimes increased when Vdd is decreased, sometimes it decreases). However, if the clock skew in 130 nm is overestimated (to be the same as the delay of 7 inverters in 130 nm technology, for example), the same trend of Fig. 9.9 can be observed in 130 nm.

This confirms that the decrease in Vdd can decrease the probability of violation, but this decrease is not so significant (only about two times) as in the other cases.

9.3.4 Dependency on Padding

The following analysis was about the padding of logic paths. Padding is the placement of extra delay in the fast logic paths to increase the race immunity. This can be done by adding extra inverters, buffers, or any other delay element.

This analysis considered each step of padding as the increase of the delay equivalent to the delay of one inverter of the given technology. Figure 9.12 presents the results for different scenarios.

The figure shows very consistent results. The placement of only one single extra inverter helps significantly to the decrease in the probability of violation. This decrease can be 100 times or up to more than 1 million times.

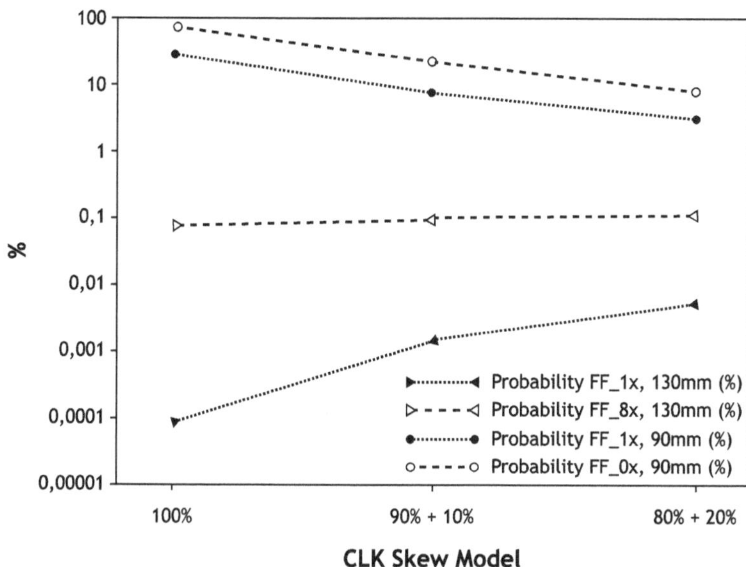

Fig. 9.13 Probability of violation for different clock skew models

This confirms that the padding can be an excellent technique to protect digital circuits against hold time violations, no matter the technology, FF or supply voltage.

9.3.5 Dependency on Clock Skew Model

The last analysis was about the dependence of probability when using different clock skew models in the calculation. The models were the ones described previously: clock skew is a fixed value in all cases (100 % of the worst case skew found in literature); the clock skew is a normal Gaussian distribution, where the average is 90 % of the worst case, and the 3 sigma standard deviation is 10 % of the worst case; and a third case where the clock skew is again a normal Gaussian distribution, but average and 3 sigma are 80 % and 20 % of the worst case, respectively. Figure 9.13 presents the results for different scenarios.

The results do not have always the same tendency, but can be explained. When the probability is high in the 100 % case, when the distributions are considered, the probability decreases significantly (about 10 times in the two 90 nm cases). This is expected, since the clock skew average is smaller in the other cases. However, when the probability is very small (like in the FF_1x, 130 nm case), the probability drastically increases (almost 100 times). This is because that the 0.3 % of cases that are outside the 3 sigma distribution will play a major role in the total probability. Finally, when the probability is in a somewhat middle value, the two effects cancel each other, and the probability remains almost stable between the different clock skew models.

Chapter 10
Protecting Circuits Against Hold Time Violations

Due to Process Variability

In this chapter, we show how to protect digital circuits against hold time violations due to process variability. First, a motivation in this issue is drawn. Then different options of how to provide the protection are presented.

10.1 Motivation

In Chap. 2, we discussed the sources of process variations and its impact on circuit level design. It is known that process parameter variability produces a statistical variation in the delay of the circuits, making some circuits faster and others slower than the nominal case. Sometimes, the circuits may be too slow or too leaky to be appropriate for sale, reducing the profit. This is called parametric yield.

Although it may not necessarily be a critical problem, variability in logic delay must be analyzed in the context of circuit design. The circuits usually have 10s or even 100s of critical paths (or with delay close to the critical one). If only one of these critical paths is slower than the nominal value, the whole circuit will have to operate at a slower clock to work properly, decreasing overall performance. As the number of critical paths increases, the probability that it happens increases, achieving high values for high performance circuits in state of the art technologies. This is currently the main topic of variability research, how to not have this big loss of performance due to delay variations. However, performance penalty is not the only issue.

Chapter 3 showed the basic aspects of FF operation. There are important characteristics to remember: delay, setup time, hold time, race immunity. And they are also subject to variations. Moreover, they can change drastically for different FF types.

Setup time is closely related with the critical paths. A long path that is slower than expected will make the circuit present a setup time violation. However, decreasing the clock frequency can make the circuit work again.

On the other hand, hold time violations are related with short paths that operate faster than nominal, making a FF input change too fast. And in this case, at nominal

G. Neuberger et al., *Protecting Chips Against Hold Time Violations Due to Variability*,
DOI 10.1007/978-94-007-2427-3_10,

supply voltage and temperature, there is nothing that can be done to make the circuit to work properly. The clock frequency does not change hold time violations. The error is permanent (at least at same temperature and Vdd).

The main recent researches in this field are about setup time violations and critical paths, since it decreases performance and profit. [49, 50] present EinsStat, a statistical timing analysis tool developed at IBM that makes the statistical analysis of critical paths and paths close to the critical ones to adjust and optimize the paths for the minimum clock period to avoid setup time violations and perform yield prediction. [6] also presents a set of CAD tools developed at University of Michigan to cope with process variability, maximizing yield and frequency, analyzing the most critical paths. However, none of these works address the very short paths subject to hold time violations.

So, there is a need for the research on hold time violations due to process variations. Due to the development of new process technologies, we consider that it can be an even worse problem, because of the following reasons: the error is permanent, while setup time violations can be corrected decreasing the clock speed; and while there are 10s or 100s of critical paths that cause setup time violations, there are at least 1,000s or 10,000s short paths that can lead to hold time violations. These all can lead to fabricated circuits unable to operate at any frequency, resulting in a very low yield.

The protection of digital circuits against hold time violations is an increasingly important topic and needs to be considered in circuit design. The next section presents different ways to cope with this problem, based on the results achieved in previous chapters.

10.2 Protection Against Hold Time Violations

In Chap. 9 the increasing probability of hold time violations was studied in different scenarios. Now we discuss different ways to protect digital circuits against these hold time violations.

The problem of hold time violations (race conditions) has been known in the industry already (although without the role of process variations), and different empirical ways to cope with it were proposed: vdd reduction, race immunity increase, and insertion of delay elements (padding). Here we look into these techniques and put together the process variability influence.

10.2.1 Vdd Reduction

It is reported in the industry that decreasing the operating voltage of a circuit that is not working due to race conditions, the circuit sometimes may work in the new voltage. This was first checked empirically, but the results from previous chapters allow us to explain this behavior.

Figure 9.11 showed the dependence of violation probability on Vdd at 90 nm technology. The probability decreased about half of its value when Vdd was decreased from 1.32 to 0.9 V. This means that in this case, half of the short paths presenting the violation may come back to work when the Vdd is decreased. If the circuit is not working due to only one or a few short paths, some of the dies may work at a lower Vdd. However, this technique is very limited, since there is not a very significant decrease in the violation probability. It is recommended only as a last attempt when the circuit was already fabricated and some dies are not working. For circuits under design, it is recommended to use techniques that allow changes and protection before fabrication, like the ones discussed below.

This effect happens because, although the variability increases under lower Vdd, the clock skew increases less significantly than the race immunity (and then the difference between race immunity and clock skew, characterizing a hold time violation, also increases).

It is important to note that these results were also achieved in our experiments: always that the Vdd was decreased in the same die, the critical clock skew changed and the circuit under test was back to function; it was needed to turn on more capacitances to produce a hold time violation again, with a higher critical clock skew.

10.2.2 Increasing Flip-Flop Race Immunity

When the FF race immunity increases, the probability of violation may decrease significantly, as shown in Figs. 9.9 and 9.10. However, this improvement is very inconstant: for a 12 % increase in race immunity, it may provide only about two times of improvement (as in 90 nm) or up to 1,000 times (as in a case in 130 nm technology).

This technique is very limited, since the fast paths subject to hold time violations probably already have slower FFs with high race immunity, since they are not critical paths. But it is always recommended to check if the synthesis tool is properly programmed.

Another point is that FFs with test inputs for scan chains have higher race immunities, since a pass transistor precedes the data input. A use of these FFs may help to prevent hold time violations in some cases, but they have a significant area overhead, and then will be a worse option than the next techniques (padding) in these cases.

Because of these points, this work will not discuss this technique in detail, and it will focus more in the most effective way to prevent against hold time violations: the insertion of extra delay in fast paths (padding).

10.2.3 Padding

The padding is the most effective way to prevent digital circuits against hold time violations, as shown in Fig. 9.12.

Padding was already presented as a technique to prevent against hold time violations in short paths by other authors [41]. However, it was considered only that the race immunity must be higher than the clock skew, without taking the role of variability into consideration. In this case, the amount of delay that padding must insert in a given path can be easily calculated as (Eq. 10.1):

$$\text{Delay\_Padding} = \text{clk\_skew} - \text{R} \tag{10.1}$$

where R is the FF race immunity.

Usually the padding is made by inserting delay elements (such as buffers), thus the delays that can be used are discrete. In this way, the number of delay elements that must be used can be easily calculated (Eq. 10.2):

$$\text{Number of delay elements} = \frac{\text{clk\_skew} - \text{R}}{\text{delay\_element}} \tag{10.2}$$

However, to take into account race immunity variability due to process variations, several changes must be made. The delay that must be inserted in the paths is not a number, but a probability function.

In the statistical framework, it is not possible to assure that the paths will work with 100 % certainty; we need to assume a threshold in the probability of hold time violations that we want. This threshold must be enough to guarantee that the circuit yield will be high enough. It depends on the number of paths in the circuit. For example, this threshold can be like 0.001 %. It means that the probability of a single path fails due to a hold time violation must be less than 0.001 %. In current technologies, this would be enough to achieve a high yield.

Taking this into consideration, an algorithm to perform the padding in the circuit was formulated. Figure 10.1 shows this algorithm.

The inputs of the algorithm are the clock skew expected in the circuit, the threshold of probability required, and the delay of each delay element that can be inserted in the circuit. The algorithm makes use of the error function to calculate the probability of violation, while increasing the number of the delay elements until the probability is smaller than the threshold.

This algorithm is considering the clock skew as a fixed, worst case value. For a more realistic scenario, where the clock skew is modeled also as a normal distribution, with average and standard deviation, the only modification needed in the algorithm is the change in the error function, to consider Eq. 9.3 instead of Eq. 9.2.

```
algorithm doPadding(clk_skew, prob_threshold, delay_element)
{
// run for all paths in the circuit
for (all paths)
  {
  // get race immunity average and standard deviation of the current path
  double R = getRaceImmunityAverage(current_path);
  double sR = getRaceImmunityStandardDeviation(current_path);
  // reset the number of delay elements needed in current path
  int num_delay_elements = 0;
  // calculate the probability of hold time violation
  double prob = erf((clk_skew – R) / (sR * sqrt(2)))/2 + 1/2;
  // repeat while the probability is not smaller than the desired probability
threshold
  while (prob > prob_threshold)
    {
    // increase the number of delay elements
    num_delay_elements++;
    // calculate the new probability
    prob = erf((clk_skew – R – num_delay_elements * delay_element) /
(sR*sqrt(2)))/2 + 1/2;
    }
  // insert the delay elements in the netlist to protect the current path
  insertInNetlist(current_path, num_delay_elements);
  }
}
```

Fig. 10.1 Algorithm of padding considering race immunity variability

Chapter 11
Padding Efficiency

Of the Proposed Padding Algorithm

To evaluate the algorithm proposed earlier in Chap. 10, a set of experiments were performed, comparing the proposed algorithm to other methods to protect against hold time violations.

11.1 Methodology

Three different scenarios were compared: the Scenario 1 is without any protection, where no elements are inserted into the short paths; Scenario 2 is inserting the amount of delay elements needed to achieve a worst-case protection of 3-sigma in both race immunity (R) and clock skew; and Scenario 3 is using the proposed algorithm.

However, after the first experiments, it was clear that the 3-sigma tolerance was not enough to guarantee a correct functioning of a large circuit, since the number of delay elements inserted by this method was even smaller than by our algorithm, and the fault probability would be sometimes even higher than 10%. So, the Scenario 2 was divided in different cases: using 3-, 4-, 5-, and 6-sigma tolerance, in order to allow a comprehensive comparison.

The experiments were performed in the following way. First, values for R and CLK_Skew were picked, in a way to consider all combinations of small or large values for both variables. A variability of sigma = 5 % for R and sigma = 10 % for CLK_Skew were considered in all cases. Then the probability of violation for one FF is calculated, and the cumulative fault probability for a circuit containing 10000 FFs is estimated. In Scenario 3, delay elements are inserted until the cumulative fault probability is smaller than 0.1 %. Three different sizes of delay elements are tested: 10 ps (which could be achieved by a small buffer in a sub-100 nm technology), 5 and 1 ps (which could be achieved with small changes in the parasitic capacitances in the already existent cells).

G. Neuberger et al., *Protecting Chips Against Hold Time Violations Due to Variability*, 97
DOI 10.1007/978-94-007-2427-3_11, © Springer Science+Business Media Dordrecht 2014

11.2 Results

Table 11.1 shows the results using a race immunity value of 150 ps and a clock skew value of 100 ps, which would be typical values for these parameters, close to the values obtained in experimental results.

From the results, it is possible to observe that with these values of race immunity and clock skew, the use of no protection leads to a high probability of failure, with 27 % of all fabricated circuits failing solely due to hold time violation. Even using a protection of 3-sigma, the probability could still be about 10 % (when delay elements of 1ps are considered). The number of delay elements inserted by our algorithm was closer to the 4-sigma protection case than to the other cases. The use of protection equivalent to 5-sigma or 6-sigma protection would provide even higher and maybe unnecessary protection; with the cost of using too many delay elements.

The next experiment was performed with the same race immunity value than the previous experiment (150 ps), but with a smaller clock skew, with the value of 80 ps, considering for a case where the clock skew is better controlled. Table 11.2 shows the results for this case.

In this experiment, the race immunity was already much larger than the clock skew, and variability could not make the fault probability significant even with no protection at all. So, most scenarios had the same result: the use of delay elements is not necessary. Only with the extreme cases of 5- and 6-sigma, delay elements are necessary.

The next experiment is using both race immunity and clock skew with smaller values than previously: 70 ps and 50 ps, respectively. Table 11.3 presents the results for this case.

Table 11.1 Results for R = 150 ps, CLK_Skew = 100 ps

Delay element	Padding		Nr Delay elements	Fault probability (%)
D = 10 ps	Scenario 1		0	27.1463416771
	Scenario 2	3σ	1	0.7901899302
		4σ	2	0.0107170160
		5σ	4	0.0000003011
		6σ	6	0.0000000000
	Scenario 3		2	0.0107170160
D = 5 ps	Scenario 1		0	27.1463416771
	Scenario 2	3σ	1	5.2686869954
		4σ	4	0.0107170160
		5σ	8	0.0000003011
		6σ	11	0.0000000000
	Scenario 3		3	0.0995946401
D = 1 ps	Scenario 1		0	27.1463416771
	Scenario 2	3σ	3	10.5741609690
		4σ	20	0.0107170160
		5σ	38	0.0000009612
		6σ	55	0.0000000000
	Scenario 3		15	0.0995946401

Table 11.2 Results for R = 150 ps, CLK_Skew = 80 ps

Delay element	Padding		Nr Delay elements	Fault probability (%)
D = 10 ps	Scenario 1		0	0.0000865707
	Scenario 2	3σ	0	0.0000865707
		4σ	0	0.0000865707
		5σ	1	0.0000001490
		6σ	3	0.0000000000
	Scenario 3		0	0.0000865707
D = 5 ps	Scenario 1		0	0.0000865707
	Scenario 2	3σ	0	0.0000865707
		4σ	0	0.0000865707
		5σ	2	0.0000001490
		6σ	5	0.0000000000
	Scenario 3		0	0.0000865707
D = 1 ps	Scenario 1		0	0.0000865707
	Scenario 2	3σ	0	0.0000865707
		4σ	0	0.0000865707
		5σ	8	0.0000005678
		6σ	23	0.0000000000
	Scenario 3		0	0.0000865707

Table 11.3 Results for R = 70 ps, CLK_Skew = 50 ps

Delay element	Padding		Nr delay elements	Fault probability (%)
D = 10 ps	Scenario 1		0	99.4744812044
	Scenario 2	3σ	1	0.4420461983
		4σ	2	0.0000280351
		5σ	3	0.0000000000
		6σ	4	0.0000000000
	Scenario 3		2	0.0000280351
D = 5ps	Scenario 1		0	99.4744812044
	Scenario 2	3σ	2	0.4420461983
		4σ	3	0.0048863419
		5σ	5	0.0000000834
		6σ	7	0.0000000000
	Scenario 3		3	0.0048863419
D = 1 ps	Scenario 1		0	99.4744812044
	Scenario 2	3σ	6	9.7161181368
		4σ	14	0.0126794799
		5σ	23	0.0000009247
		6σ	31	0.0000000000
	Scenario 3		12	0.0789260167

The results were close to the ones in the first experiment, since both values are scaled, but now the use of no protection leads to an almost 100 % chance of failure. Again, the results of our algorithm are very close to the 4-sigma case.

The next case is with a larger clock skew (100 ps), but the same race immunity remains the same (70 ps). Table 11.4 shows these results.

Table 11.4 Results for R = 70 ps, CLK_Skew = 100 ps

Delay element	Padding		Nr delay elements	Fault probability (%)
D = 10 ps	Scenario 1		0	99.9999999915
	Scenario 2	3σ	8	1.1763625819
		4σ	9	0.0074306747
		5σ	10	0.0000196082
		6σ	12	0.0000000000
	Scenario 3		9	0.0074306747
D = 5 ps	Scenario 1		0	99.9999999915
	Scenario 2	3σ	15	10.2508339885
		4σ	17	0.1044052219
		5σ	20	0.0000196082
		6σ	23	0.0000000000
	Scenario 3		18	0.0074306747
D = 1 ps	Scenario 1		0	99.9999999915
	Scenario 2	3σ	71	41.9923436096
		4σ	84	0.1725404345
		5σ	98	0.0000689308
		6σ	111	0.0000000104
	Scenario 3		86	0.0626247612

Due to the fact that now the clock skew is already larger than the race immunity, many delay elements are needed to protect the circuit in all scenarios. Using only 3-sigma protection, the fault probability can be up to 42 %. With no protection at all, almost all circuits will fail. The proposed algorithm successfully provides the number of delay elements to protect the circuit in an efficient way.

After several combinations of values of race immunity and clock skew were tested, a different experiment was performed: considering a different number of FFs that could fail in the circuit. First, a small circuit with 1000 FFs instead of 10000 FFs is investigated. Later, a larger circuit with 100000 FFs is tested. For these experiments, we will return to the first scenario, with a race immunity of 150 ps, and a clock skew of 100ps, where we could compare with the results previously achieved.

Table 11.5 Presents the Results for the Case with 1000 FFs. Comparing these results with the ones previously presented in Table 11.1, it is possible to see that even using no protection, and the fault probability is decreased to 3 %, compared to 27 % in the previous case. And now, a 3-sigma protection provides enough protection, with 4-sigma being overwhelming, using unnecessary delay elements.

In this case, it is possible to see that our algorithm provides the best solution, using an intermediary result between 3- and 4-sigma, even when these parameters of worst-case tolerance are not known to the algorithm. Finally, Table 11.6 shows the results using a circuit with 100000 FFs.

In this case, no protection got a probability even higher, with 95 % of fault probability, and 3-sigma is also an inappropriate protection, with a high (67 %)

Table 11.5 Results for R = 150 ps, CLK_Skew = 100 ps, Number of FFs=1000

Delay element	Padding		Nr delay elements	Fault probability (%)
D = 10 ps	Scenario 1		0	3.1175447388
	Scenario 2	3σ	1	0.0793013874
		4σ	2	0.0010717533
		5σ	4	0.0000000301
		6σ	6	0.0000000000
	Scenario 3		1	0.0793013874
D = 5 ps	Scenario 1		0	3.1175447388
	Scenario 2	3σ	1	0.5397937067
		4σ	4	0.0010717533
		5σ	8	0.0000000301
		6σ	11	0.0000000000
	Scenario 3		2	0.0793013874
D = 1 ps	Scenario 1		0	3.1175447388
	Scenario 2	3σ	3	1.1113831778
		4σ	20	0.0010717533
		5σ	38	0.0000000961
		6σ	55	0.0000000000
	Scenario 3		10	0.0793013874

Table 11.6 Results for R = 150 ps, CLK_Skew = 100 ps, Number of FFs=100000

Delay element	Padding		Nr Delay elements	Fault probability (%)
D = 10 ps	Scenario 1		0	95.7877547802
	Scenario 2	3σ	1	7.6267588792
		4σ	2	0.1071184903
		5σ	4	0.0000030109
		6σ	6	0.0000000000
	Scenario 3		3	0.0007768822
D = 5 ps	Scenario 1		0	95.7877547802
	Scenario 2	3σ	1	41.7983136755
		4σ	4	0.1071184903
		5σ	8	0.0000030109
		6σ	11	0.0000000000
	Scenario 3		5	0.0098653904
D = 1 ps	Scenario 1		0	95.7877547802
	Scenario 2	3σ	3	67.2937888850
		4σ	20	0.1071184903
		5σ	38	0.0000096123
		6σ	55	0.0000000000
	Scenario 3		21	0.0673246981

chance of failure. The 4-sigma scenario leads to a probability slightly higher than 0.1 %, which is the threshold used in the proposed algorithm. Because of this, the algorithm uses more delay elements than 4-sigma, providing again the best balance between the fault probability and the number of delay elements used.

Another point that shows the strength of the proposed algorithm is that it can easily handle the correlation between the FFs. The necessary calculation can be easily provided in the step of calculating the cumulated fault probability based on the individual FF violation probability. The cases in Scenario 2 cannot handle correlation efficiently, since it cannot infer a fault probability in standard deviations (σ's) with correlation for different sizes of circuits.

Chapter 12
Final Remarks

The current IC technologies are subject to an increasingly sensitivity to process variations. These variations affect the design of digital circuit in different ways. The focus of this book is the effect of variability on hold time violations.

This work started by reviewing process variability, flip-flops, and hold time violations. Then by the use of simulations, the increasingly importance of hold time violations due to process variability was verified. This is a very important topic, because once a circuit with hold time violations is fabricated, almost nothing can be made to make it function again.

After the detailed study of the hold time violation mechanism by circuit level Monte Carlo simulation, the test circuit for detailed experimental study of the topic is designed. For accurate on-wafer characterization, a test circuit and a measurement technique with ~1 ps resolution are presented. This measurement technique can be also applied to accurately measure other characteristic of the circuit, such as setup time, and is one of the main contributions of this book.

Using this technique, this work presents an experimental analysis of the variability of race immunity of edge-triggered master–slave FFs due to process variations in 130 and 90 nm low power CMOS technologies. The presented methodology provides detailed information on circuit robustness of FFs under realistic operating conditions.

A wide range of experimental results is presented, such as for different technologies (130 and 90 nm), different FFs, different operating temperatures, and under different supply voltages. This provided a wide and accurate characterization for the next steps of this work.

The difference between systematic and random residual parameter variations is also studied. The raw measured data includes both components, but using numerical methods, it is possible to separate them. Using these numerical separation methods, the data for both systematic and random residual variations are unveiled. This separation is important, since there may be different ways to cope with them at the circuit and process levels. The results showed that, while systematic variations are decreasing in 90 nm due to better process control, the random residual variations

G. Neuberger et al., *Protecting Chips Against Hold Time Violations Due to Variability*, DOI 10.1007/978-94-007-2427-3_12, © Springer Science+Business Media Dordrecht 2014

are increasing, since they are more difficult to control and increase naturally in smaller dimensions, due to the stochastic nature of the process involved, as for instance random dopant fluctuations.

After the separation between systematic and random variability, it was necessary to apply different normality tests in the data, to check if they are normal Gaussian distributions. The results were that in 130 nm the systematic variability is not normal Gaussian, while the random residual is. In 90 nm, both of them were normal Gaussian distributions, with results more significant than in 130 nm.

Since hold time violations are also closely related to clock skew, it is also very important to analyze the case when both race immunity and clock skew become random variables. The probability of hold time violations would be very small if the clock skew is much smaller than the race immunity. However, works in the literature show that the order of magnitude of clock skew is close to the one of race immunity. These numbers were used to calculate the probability of violation under different scenarios. The results show that a small increase of the race immunity can decrease significantly the probability of violation; the reduction of Vdd has some impact to decrease the probability of violation; the insertion of delay elements (padding) have a very large impact (sometimes decreasing the violation probability up to one million times); and more importantly, in the more recent technologies there is a dramatic increase in the probability of violation, meaning that this problem will become of ever increasing relevance for future technologies. The effect of considering different clock skew models was also analyzed. When the clock skew is considered to be a random variable, with given average and a standard deviation, the probability decreases if it was too high in the worst case value; however it decreases if in the worst case it was too low.

Finally, different ways to protect digital circuits against hold time violations were presented, being the most important of them the use of padding. An algorithm for padding was proposed with adjustments to take into account the role of race immunity variability, thus reducing significantly the probability of hold time violations, and increasing the yield.

References

1. ANDERSON, T. W.; DARLING, D. A. Asymptotic Theory of Certain "Goodness-of-Fit" Criteria Based on Stochastic Processes. Annals of Mathematical Statistics, Ann Arbor, v. 23, p. 193–212, 1952.
2. ASHAN, I. et al. RTAdriven Intra-die Variations in Stage delay and Parametric Sensitivities for 65 nm Technology, in VLSI Symp. Tech. Dig., Honolulu, HI, 2006, pp. 170–171.
3. ASENOV, A. Personal Communication. Work presented at 3rd European Workshop on Ultimate Integration of Silicon – ULIS, 2002, Munich, Germany. Workshop without formal proceedings.
4. AUSSCHNITT, C. et al. Industrial Strength Lithography APC. In: AMM Conference, 2003. Proceedings... [S.l.: s.n.], 2003. p. 1–11.
5. BALSARA, P. T.; KO, U. High-Performance Energy-Efficient D-Flip-Flop Circuits. IEEE Transactions on VLSI Systems, New York, v. 8, p. 94–98, Feb. 2000.
6. BLAAUW, D.; CHOPRA, K. CAD Tools for Variation Tolerance. In: Design Automation Conference, DAC, 42., 2005. Proceedings... New York: IEEE/ACM, 2005.
7. BONING, D. S.; CHUNG, J. E. Statistical metrology: understanding spatial variation in semiconductor manufacturing. In: Microelectronic Manufacturing Yield, Reliability and Failure Analysis II. Proceedings... Bellingham, Washington: SPIE, 2006. p. 16–26.
8. BORKAR, S. Designing reliable systems from unreliable components: the challenges of transistor variability and degradation. IEEE Micro, New York, v. 25, n. 6, p. 10–16, Nov./Dec. 2005.
9. BRUNNER, T. A. et al. High NA Swing Curve Effects. In: Optical Microlithography, 14., 2001, Santa Clara, CA. Proceedings... [S.l.]: SPIE, 2001. p. 1050–1057.
10. BRUNNER, T. A. Why Optical Lithography Will Live Forever. Journal of Vaccum Science & Technology B: Microelectronics and Nanometer: processing, measurement, and phenomena, New York, v.21, n.6, p. 2632–2637, 2003.
11. DALLY, W.; POULTON, J. Digital Systems Engineering. Cambridge, UK: Cambridge University Press, 1998.
12. DAO, H. Q.; NOWKA, K.; OKLOBDZIJA, V. G. Analysis of Clocked Timing Elements for Dynamic Voltage Scaling Effects over Process Parameter Variation. International Symposium on Low Power Electronics and Design, ISLPED, 2001. Proceedings... New York: ACM, 2001. p. 56–59.
13. EDIRISOORIYA, S.; EDIRISOORIYA, G. Diagnosis of Scan Failures. In: VLSI Test Symposium, 1995. Proceedings... [S.l.: s.n.], 1995. p. 250–255.
14. FINCH, S. J.; THODE, H. Testing for Normality. [S.l.]: M. Dekker, 2001.
15. FRANK, D. J.; WONG, H.-S. P. Simulation of Stochastic Doping Effects in Si MOSFETs. In: International Workshop on Computational Electronics, IWCE, 7., 2000. Proceedings... Glasgow: University of Glasgow, 2000.

G. Neuberger et al., *Protecting Chips Against Hold Time Violations Due to Variability*, DOI 10.1007/978-94-007-2427-3,

16. FRANK, D. J. Power-Constrained CMOS Scaling Limits. IBM Journal on Res. Devel., Armonk, v. 46, n. 2/3, Mar./May 2002.
17. FRANK, D. J. Parameter Variations in Sub-100 nm CMOS Technology. In: Advanced Solid State Circuits Forum, ISSCC, 2004. Proceedings… [S.l.: s.n.], 2004.
18. FRIEDBERG, P.; CHEUNG, W.; SPANOS, C. J. Spatial Modeling of Micron-Scale Gate Length Variation. In: Data Analysis and Modeling for Process Control III, 2006. Proceedings… [S.l.]: SPIE, 2006. p. 61550C.1-61550C.12.
19. GEROSA, G. et al. A 2.2W, 80 MHz Superscalar RISC Microprocessor. IEEE Journal of Solid-State Circuits, New York, v. 29, n. 12, p. 1440–1454, Dec. 1994.
20. GUO, R.; VENKATARAMAN, S. A Technique for Fault Diagnosis of Defects in Scan Chains. In: International Test Conference, 2001. Proceedings… [S.l.: s.n.], 2001. p. 268–277.
21. HARRIS, D.; SUTHERLAND, I.; SPROUL, B. Logical Effort: Designing Fast CMOS Circuits. San Francisco, CA: Morgan Kaufmann, 1999.
22. HODGES, D.; JACKSON, H. Analysis and Design of Digital Integrated Circuits. New York: McGraw Hill, 1988.
23. HUANG, Y. et al. Statistical Diagnosis for Intermittent Scan Chain Hold-Time Fault. In: International Test Conference, 2003. Proceedings… [S.l.: s.n.], 2003. p. 319–328.
24. JOANES, D. N.; GILL, C. A. Comparing Measures of Sample Skewness and Kurtosis. Journal of the Royal Statistical Society (Series D): The Statistician, Oxford, v. 47, n. 1, p. 183–189, 1998.
25. KUHN, K.; KENYON, C.; KORNFELD, A.; LIU, M.; MAHESHWARI, A.; SHIW, W.; SIVAKUMAR, S.; TAYLOR, G.; VANderVOORN, P.; ZAWADZKI, K. Managing Process Variation in Intel's 45nm CMOS Technology. Intel Technology Journal, v. 12, n. 2, p. 93–110, 2008.
26. LI, J. C.-M. Diagnosis of Multiple Hold-Time and Setup-Time Faults in Scan Chains. IEEE Transactions on Computers, Los Alamitos, v. 54, n. 11, p. 1467–1472, Nov. 2005.
27. LIEBMANN, L. W. et al. TCAD Development for Lithography Resolution Enhancement, IBM J. Res. Develop., vol. 45, no. 5, pp. 651–665, Sep. 2001.
28. MACH, P.; HOCHLOVA, H. Testing of Normality of Data Files for Application of SPC Tools. In: International Spring Seminar on Electronics Technology: Meeting the Challenges of Electronics Technology Progress, 27., 2004. Proceedings… [S.l.]: IEEE, 2004. v. 2, p. 318–321.
29. Marković, D. Analysis and Design of Low-Energy Clocked Storage Elements. 2000. Master Dissertation, Electrical Engineering and Computer Sciences – University of California at Berkeley.
30. Marković, D.; Nikolić, B.; BRODERSEN, R. W. Analysis and Design of Low-Energy Flip-Flops. In: International Symposium on Low Power Electronics and Design, ISLPED, 2001. Proceedings… New York: ACM Press, 2001. p. 52–55.
31. NAJM, F. N. "On the Need for Statistical Timing Analysis", Proc. Design Automation Conference (DAC), pp. 764–765, 2005.
32. MEHROTRA, V.; BONING, D. Technology Scaling Impact of Variation on Clock Skew and Interconnect Delay. In: IEEE International Interconnect Technology Conference, 2001. Proceedings… [S.l.]: IEEE, 2001. p. 122–124.
33. NEUBERGER, G.; KASTENSMIDT, F.; REIS, R.; WIRTH, G.; BREDERLOW, R.; PACHA, C. Statistical Characterization of Hold Time Violations in 130nm CMOS Technology. In: European Solid-State Circuits Conference, 32., 2006, Montreux. Proceedings… Piscataway: IEEE, 2006. p. 114–117.
34. NEUBERGER, G.; KASTENSMIDT, F.; WIRTH, G.; REIS, R.; BREDERLOW, R.; PACHA, C. Statistical Analysis of Variability of Flip-Flop Race Immunity in 130nm and 90nm CMOS Technologies. In: Latin-American Test Workshop, LATW, 8., 2007, Cuzco, Peru. Proceedings… [S.l.: s.n.], 2007.
35. NIST. NIST/Sematech Dataplot Databook. United States, June 2001. Available at: <http://www.itl.nist.gov/div898/software/dataplot/homepage.htm>. Visited on: March 2007.
36. RABAEY, J. M. Digital Integrated Circuits: A Design Perspective. Upper Saddle River, NJ: Prentice Hall, 1996.

37. ROHRER, N. J. Introduction to Statistical Variation and Techniques for Design Optimization. Personal Communication, Tutorial presented at International Solid-State Circuits Conference, ISSCC, 2006.
38. SALEH, R. et al. Clock Skew Verification in the Presence of IR-Drop in the Power Distribution Network. IEEE Transactions on Computer-Aided Design of Integrated Circuits and Systems, v. 9, n. 6, p. 635–644, June 2000.
39. SAXENA, S. et al. Variation in Transistor Performance and Leakage in Nanometer-Scale Technologies", IEEE Transactions on Electron Devices, Jan. 2008, Vol 55 Issue 1, pp. 131–144.
40. SHAPIRO, S. S.; WILK, M. B. An Analysis of Variance Test for Normality (Complete Samples). Biometrika, [S.l.], v. 52, n. 3–4, p. 591–611, 1965.
41. SHENOY, N. V.; BRAYTON, R. K.; SANGIOVANNI-VINCENTELLI, A. L. Minimum Padding to Satisfy Short Path Constraints. In: IEEE/ACM International Conference on Computer Aided Design, ICCAD. Proceedings… [S.l.: s.n.], 1993. p. 156–161.
42. SOONG, T. T. Fundamentals of Probability and Statistics for Engineers. [S.l.]: John Wiley & Sons, 2004.
43. SPRINGER, S. K. et al. Modeling of Variation in Submicrometer CMOS ULSI technologies, IEEE Trans. Electron Devices, vol. 53, no. 9, pp. 2168–2178, Sep. 2006.
44. STEPHENS, M. A. EDF Statistics for Goodness of Fit and Some Comparisons. Journal of the American Statistical Association, Washingtion, v. 69, p. 730–737, 1974.
45. STINE, B. E.; BONING, D. S.; CHUNG, J. E. Analysis and Decomposition of Spatial Variation in Integrated Circuit Processes and Devices. IEEE Transactions on Semiconductor Manufacturing, New York, v. 10, n. 1, p. 24–41, Feb. 1997.
46. STOJANOVIC, V. OKLOBDZIJA, V. G. Comparative Analysis of Máster-Slave Latches and Flip-Flops for High-Performance and Low-Power Systems. IEEE Journal of Solid-State Circuits, New York, v. 34, p. 536–548, Apr. 1999.
47. SVENSSON, C.; YUAN, J. Latches and Flip-Flops for Low Power Systems. In: CHANDRAKASAN, A., BRODERSEN, R. (Ed.). Low Power CMOS Design. Piscataway, NJ: IEEE Press, 1998. p. 233–238.
48. VAN DER LANN, H. et al. "Etch, Reticle, and Track CD Fingerprint Corrections with Local Dose Compensation", Proc. SPIE, vol. 5755, pp. 107–118, 2005.
49. Visweswariah, C. Statistical Analysis and Design of Digital Integrated Circuits. Invited Presentation at EDA Forum, 2005, Hannover, Germany.
50. Visweswariah, C. Statistical Analysis and Optimization in the Presence of Gate and Interconnect Delay Variations. In: International Workshop on System Level Interconnect Prediction, SLIP, 2006, Munich. Proceedings… New York: ACM, 2006. p. 37.
51. WANG, Z. et al. Diagnosis of Hold Time Defects. In: IEEE International Conference on Computer Design: VLSI in Computer and Processors, ICCD, 2004. Proceedings… [S.l.]: IEEE, 2004. p. 192–199.
52. WONG, A. K. Resolution Enhancement Techniques in Optical Lithography. Bellingham, WA: SPIE Press, 2001.
53. ZARKESH-HA, P.; MULE, T.; MEINDL, J. D. Characterization and Modeling of Clock Skew with Process Variations. In: IEEE Custom Integrated Circuits, 1999. Proceedings… [S.l.]: IEEE, 1999. p. 441–444.

Zeitfracht Medien GmbH
Ferdinand-Jühlke-Straße 7
99095 Erfurt, Deutschland
produktsicherheit@kolibri360.de